Integrated Optics

in silicon-based aluminum oxide

The photograph on the cover shows the mask pattern of a wavelength demultiplexer, based on an optical phased array. (see also figure 6.34a)

Integrated Optics

in silicon-based aluminum oxide

Proefschrift

ter verkrijging van de graad van
doctor
aan de Technische Universiteit Delft,
op gezag van de rector magnificus,
prof. drs. P.A. Schenck,
in het openbaar te verdedigen ten overstaan van een commissie
aangewezen door het College van Dekanen
op 18 juni 1991 te 16.00 uur.

door

Meint Koert Smit

geboren te Vlissingen,
elektrotechnisch ingenieur

Dit proefschrift is goedgekeurd door de promotor

Prof. Dr. G.A. Acket

CIP-gegevens Koninklijke Bibliotheek, Den Haag

Smit, Meint Koert

Integrated optics in silicon-based aluminum oxide
Meint Koert Smit - [S.l. : s.n.]. - Ill.
Proefschrift Delft. - Met lit. opg.
Met samenvatting in het Nederlands.
ISBN 90-9004261-X
Trefw.: geïntegreerde optica / optoëlectronica.

aan Marijke

Preface and summary

Opto-electronics is one of the most rapidly developing technologies of our age. The semiconductor laser was commercially employed within ten years of its first demonstration in 1970, and within twenty years it had revolutionized the audio market. An even greater economical impact may be expected from the application of optical technology in the telecommunication network. Within twenty years after its birth in 1970 the optical fiber has pushed aside the copper cable in the intercity network. In the coming decade it will penetrate the local network and reach the private subscriber. This will open a huge market for opto-electronic interface equipment.

The Delft University of Technology entered the optical communication scene in 1977, when professor H. Blok (Electromagnetism) introduced a course on optical waveguide theory. In 1979 the scope was broadened to optical communication technology. The new course was given jointly by professors H. Blok, L. Bordewijk (Information Transmission), and L. Krul (Microwaves). In 1981 G.A. Acket, leader of the "Physics of Microwave and Optical Devices" group at Philips Research Laboratories in Eindhoven, was appointed as part-time professor of optical communication technology, in the department of Electrical Engineering.

In addition to two courses in optical communication technology, Acket began a research project in the relatively new field of integrated optics. A decision was taken to work initially with aluminum oxide as a waveguide material, and to switch to the more challenging - but also more complicated - III-V semiconductor materials at a later stage.

My involvement in this research began at its initiation. The present thesis comprises the most important results of the research program on silicon-based integrated-optical components at the Laboratory of Telecommunication and Remote Sensing Technology, carried out in the period between 1983 and 1991. The far-reaching optical similarity between the aluminum oxide waveguide system, operating at 633 nm wavelength, and InGaAsP/InP waveguide systems operating at 1300 nm wavelength (wavelengths within the waveguides are almost equal) broadens the applicability of the results reported. Recent experiments with InGaAsP/InP-based waveguide bends (not reported here) confirm this statement.

The thesis covers the following topics:

* **The development of low-loss aluminum oxide waveguide films (0.35 dB/cm @ 1300 nm wavelength)**. Fabrication process and annealing procedure are described in chapter 2. Waveguide design and fabrication are discussed in chapter 3.

* **A normalized approach to the optimal design of waveguide bends, including the junctions to straight waveguides.** In chapter 4 the normalization approach is explained and a set of empirical formulae is derived for computing the design parameters. Using them optimal bend design can be done with a pocket calculator.

* **The development of compact directional couplers with reduced sensitivity to variations in the fabrication process.** This development began with the design of conventional TMI-couplers, and has led, via two modifications introduced to improve the performance, to directional couplers based on the self-imaging properties of multimode waveguide sections (with single-mode input- and output waveguides). We called these couplers Multi-Mode Interference (MMI) couplers. Research on these couplers, which promise outstanding performance, is presently being continued by L.B. Soldano at the Delft University of Technology, and by E.C.M. Pennings at Bellcore[1]. In chapter 5 the results on TMI-couplers, including the modifications, and the fundamentals of MMI-couplers are described.

* **The development of focusing and dispersive components based on optical phased arrays for application in components like wavelength demultiplexers, filters, and polarization splitters.** Optical phased arrays form an alternative to curved planar gratings. They can be realized employing conventional (high-quality) optical lithography in the same proces step

---

1) Bellcore, Red Bank (NJ), USA.

as the waveguides, thus avoiding the additional process steps required for electron-beam or holographic lithography. Experimental results show good performance combined with small device dimensions. In chapter 6 the operation and design of optical phased arrays is described and experimental results are presented.

* **Accuracy analysis of refractive-index and film-thickness determination from ellipsometric measurement data.** Automated ellipsometers for determination of the refractive index and thickness of transparent films on (preferably) absorbing substrates are wide-spread. The dependence of their accuracy on the film-parameters is complicated. This reflects on the reliability of the automated approach. A Bounded-Error Estimation Approach has been developed and applied to the analysis of ellipsometric measurement accuracy. Results are presented in chapter 7.

* **Analysis of the efficiency of the prism-coupler for coupling light into narrow waveguide channels.** Inclusion of diffraction effects in the analysis of the coupling mechanism reveals that optimal (but low) coupling efficiencies into narrow waveguides are obtained with weakly focused beams. From a measurement point-of-view this is an interesting result. Theory and experiments are described in chapter 8. A review of measurement methods developed at the authors laboratory is included.

* **A novel method for matching strongly elliptical mode-profiles in planar waveguides to fiber-modes.** The elliptical mode profile of the planar

waveguide is matched to the fiber mode by means of a tapered interface which is produced in a simple polishing step. Theory of operation and experimental results are presented in chapter 9.

The results presented in this thesis could only be achieved through the close cooperation of many colleagues, students, and relatives. I am very grateful for their support and their contributions (a detailed account is given in the acknowledgements at the end of this thesis).

Delft, May 6, 1991

Meint K. Smit

Note: The thickness of the present thesis is partly due to the 1½ line spacing used in the manuscript. Unfortunately the available editor did not support compression of sub- and superscripts which is necessary for using single line spacing. The perceptive reader will note some minor problems with the alignment of sub- and superscript lines.

Voorwoord en samenvatting

(Preface and summary in Dutch)

Opto-elektronica is één van de snelst ontwikkelende technologieën van onze tijd. De halfgeleiderlaser werd al commercieel toegepast binnen tien jaar na zijn "geboorte" in 1970 en bracht met de compact-disc afspeelapparatuur een omwenteling teweeg op audiogebied. De toepassing van optische technologie in het toekomstige lokale telecommunicatienet zal nog grotere economische en maatschappelijke gevolgen hebben. In het interlokale net heeft de glasvezel de koperkabel nu reeds verdrongen. In het komende decennium zal dat ook in het lokale net gaan gebeuren en zal de abonnee zijn informatie via een glasvezel aangeleverd krijgen. Dit zal een enorme markt openen voor opto-elektronische schakelingen.

De Technische Universiteit Delft[1] zette de eerste stappen op het gebied van de glasvezelcommunicatie in 1977. In dat jaar startte professor H. Blok van de vakgroep Theoretische Electriciteitsleer[2] een college over de theorie van optische golfgeleiding. In 1979 werd dat college verbreed tot een college "Techniek van de glasvezelcommunicatie", dat gedoceerd werd door de hoogleraars H. Blok, L. Bordewijk (Transmissie van informatie) en L. Krul (Microgolftechniek). In 1981 werd G.A. Acket, leider van de groep "Physics of Optical and Microwave Devices" van het Natuurkundig Laboratorium van

---

1) Destijds: Technische Hogeschool Delft.
2) Thans: Electromagnetisme.

de N.V. Philips' Gloeilampenfabriek, benoemd tot deeltijdhoogleraar in de techniek van de glasvezelcommunicatie bij de vakgroep Microgolftechniek (thans Telecommunicatie en Tele-observatietechnologie - TTT) van de afdeling der Elektrotechniek.

Naast een tweetal colleges op het gebied van de glasvezelcommunicatietechniek startte Acket ook een onderzoekproject op het gebied van de geïntegreerde optica, die toen nog in de kinderschoenen stond. Als materiaal voor de optische golfgeleiders werd in eerste instantie gekozen voor aluminiumoxyde, een materiaal waarmee ruime ervaring bestond voor het coaten van laserspiegels. De overstap naar de vanuit technisch oogpunt bezien meer interessante - maar ook meer gecompliceerde - III-V halfgeleiders werd voorzien nadat eerst ervaring zou zijn opgedaan met aluminiumoxyde.

Ik had het genoegen vanaf het begin bij dit onderzoek betrokken te zijn geweest. Het voorliggende proefschrift beschrijft de belangrijkste resultaten van het onderzoekprogramma op het gebied van passieve optische componenten, zoals dat bij de vakgroep Telecommunicatie- en Teleobservatietechnologie is uitgevoerd in de periode van 1983 tot 1991. De resultaten van dit onderzoek zijn breder toepasbaar dan alleen voor aluminiumoxyde golfgeleiders. Doordat de golflengte van licht uit een He-Ne laser (633 nm) in een aluminiumoxyde golfgeleider vrijwel gelijk is aan die van licht met een vacuumgolflengte van 1300 nm (samen met 1550 nm de belangrijkste golflengten voor telecommunicatie) in een InGaAsP golfgeleider, zijn deze golfgeleidersystemen optisch bezien vrijwel identiek.

Een ontwerp dat in het ene systeem is ontwikkeld, zou daarom ook in het andere moeten werken. Recente experimenten wijzen uit dat dit inderdaad het geval is.

In het proefschrift worden de volgende onderwerpen behandeld:

* **De ontwikkeling van aluminiumoxyde golfgeleiders met geringe optische verliezen (0.35 dB/cm bij een golflengte van 1300 nm).** Het fabricageproces van de aluminiumoxydefilms en de warmtebehandeling, die nodig is om de verliezen te reduceren, worden behandeld in hoofdstuk 2. Hoofdstuk 3 behandelt het ontwerp en de vervaardiging van golfgeleiders in deze films.

* **Een genormaliseerde methode om gebogen optische golfgeleiders, met inbegrip van de aansluiting aan rechte golfgeleiders, optimaal te kunnen ontwerpen (met minimale verliezen).** De normalisatiemethode wordt uitgelegd in hoofdstuk 4. Tevens worden enkele empirische formules gegeven, die het mogelijk maken een optimale bocht te ontwerpen met behulp van een zakrekenmachine.

* **De ontwikkeling van compacte richtingskoppelaars met een geringe gevoeligheid voor variaties in het fabricageproces.** Deze onwikkeling is begonnen met het ontwerp van conventionele Two-Mode Interference (TMI) koppelaars. Via een tweetal modificaties ter verbetering van de eigenschappen, heeft zij geleid tot een nieuw type richtingskoppelaar, dat

gebaseerd is op de afbeeldende eigenschappen die multimodale golfgeleiders bezitten door het interferentiegedrag van de modi. Het onderzoek aan deze veelbelovende koppelaars, die we Multi-Mode Interference (MMI) koppelaars genoemd hebben, wordt voortgezet door L.B. Soldano bij de TU Delft en E.C.M. Pennings bij Bellcore[1]. In hoofdstuk 5 worden de resultaten behandeld die behaald zijn met conventionele en gemodifideerde TMI-koppelaars en wordt de werking van MMI-koppelaars besproken.

* **De toepassing van optische *phased arrays* voor het ontwerp van golflengtedemultiplexers, golflengtefilters en polarisatiesplitsers.** Optische *phased arrays* vormen een alternatief voor gekromde planaire tralies. Ze kunnen met behulp van conventionele optische lithografie worden gerealiseerd in hetzelfde proces als de golfgeleiders, waardoor de extra processtappen die benodigd zijn voor holografische of elektronenbundellithografie worden vermeden. Experimentele resultaten wijzen uit dat deze nieuwe componenten goede eigenschappen combineren met geringe afmetingen. In hoofdstuk 6 worden de werking en het ontwerp ervan behandeld en worden de experimentele resultaten beschreven.

* **Nauwkeurigheidsanalyse van brekingsindex- en filmdiktegegevens die verkregen zijn met behulp van ellipsometrie.** Geautomatiseerde meetopstellingen voor het bepalen van de dikte en de brekingsindex van transparante films op siliciumsubstraten zijn op tal van plaatsen in gebruik.

---

1) Bellcore, Red Bank (NJ), USA.

De nauwkeurigheid van deze methode hangt sterk af van de specifieke eigenschappen van film en substraat. Kennis hiervan is belangrijk om tot betrouwbare meetresultaten te komen. Hoofdstuk 7 geeft de resultaten van een onderzoek naar de nauwkeurigheid van deze methode met behulp van een *Bounded Error* schattingstechniek.

* **Een onderzoek naar het rendement van het inkoppelen van licht in zeer smalle golfgeleiders met behulp van de prismakoppelmethode.** Als de effecten van diffractie op het inkoppelrendement in rekening worden gebracht blijkt dat dit rendement optimaal (hoewel vrij laag) is wanneer de in te koppelen bundel niet te sterk gefocusseerd wordt. Vanuit het oogpunt van meettechniek is dit een gunstig resultaat, daar een zwak gefocusseerde bundel het inkoppelen in een golfgeleider makkelijker maakt en de gevoeligheid voor trillingen vermindert. Theorie en experimenten worden besproken in hoofdstuk 8. Tevens wordt daar een overzicht gegeven van de meetmethoden die binnen de vakgroep zijn ontwikkeld.

* **Een nieuwe methode om de sterk afgeplatte intensiteitsverdelingen in planaire golfgeleiders aan te passen aan de circulaire verdeling van een glasvezelmode.** Het afgeplatte profiel wordt omgezet in een circulair profiel, dat aangepast is aan de glasvezel, door het golfgeleideruiteinde geleidelijk in dikte te laten afnemen *(taperen)*. Dit kan gebeuren met behulp van een eenvoudige polijststap. De werking van deze methode, alsmede de experimentele resultaten, worden behandeld in hoofdstuk 9.

Het werk, dat in dit proefschrift behandeld is, kon allen tot stand komen door een intensieve samenwerking met talrijke medewerkers, zowel collega's als studenten, en steun van familieleden. Ik ben hen dankbaar voor hun bijdragen en hun steun (een gedetailleerd overzicht is te vinden in de "acknowlegements" aan het eind van dit proefschrift).

Delft, 6 mei 1991

Meint Smit

Noot: De omvang van dit proefschrift wordt deels veroorzaakt doordat het manuscript met 1½ regelafstand afgedrukt moest worden, daar de gebruikte tekstverwerker niet beschikte over de mogelijkheid "sub- en superschrift-tekst" te verkleinen. De oplettende lezer zal op enkele plaatsen problemen opmerken met de uitlijning van sub- en superschrift-regels.

Contents

Chapter 1

Introduction

This chapter sketches the history and future of telecommunication, with emphasis on the role of optical communication technology. Next the integrated-optics research at the Delft University of Technology, part of which forms the subject of the present thesis, is briefly described. A short summary completes this chapter.

1.1 Telecommunication: importance, history, and future

Telecommunication is a rapidly growing economic sector. The world market for telecommunication equipment is greater than those for consumer electronics and micro-electronic components together; in the information sector it is only exceeded by the computer market. Figure 1.1 depicts the volumes of the various sectors. Telecommunication equipment and services together make up more than half of the total information market, and are rapidly growing. They will soon exceed the market volume of the automobile industry to form the greatest market segment in the industrialized world [1].

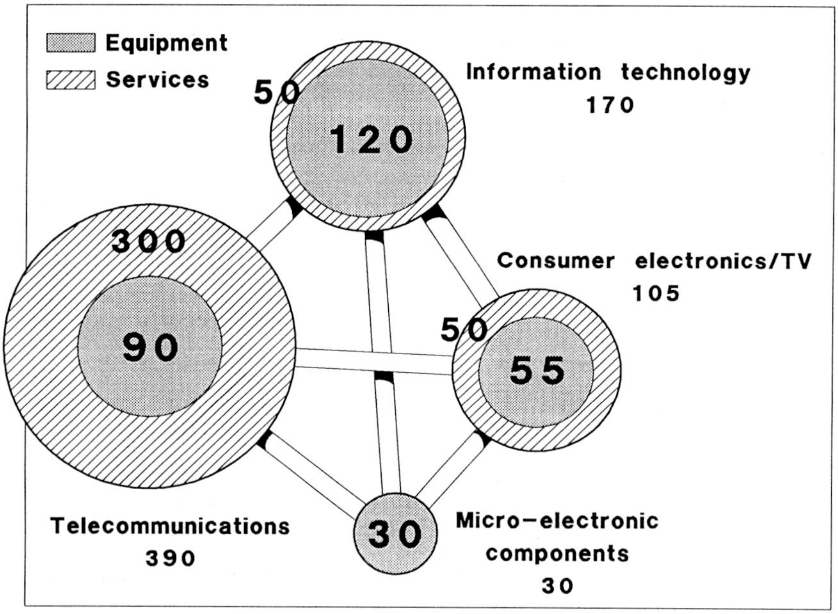

Figure 1.1 The world market for management and transmission of information in the mid-1980's (in billion ECU's, source: Studies of the Commission of the European Communities).

The oldest telecommunication event mentioned in the literature are the fires, lit on a series of successive mountains, by the ancient Greeks to herald the fall of Troy. Other examples of primitive telecommunication include the smoke signals, employed by the American Indians, and drum signals employed by numerous African peoples. These early examples of telecommunication had no sequels, however, and must therefore be viewed as historically incidental.

The roots of modern telecommunication can be traced back to the signaling systems, which were already in use in the fifteenth century for communication between ships. In the following centuries these systems, which were based mainly on the use of flags, were further developed to handle an increasing variety of messages. In 1691 Britain and the Dutch Republic agreed on a common signaling standard. A further step was the introduction of a set of flags embracing both alphabet and numbers by the British admiral von Kempenfelt in 1781. Communication by means of such a coded alphabet was called telegraphy.

Telegraphy over land began in 1767 with Edgeworth's development of the semaphore, a pole with rotatable arms. A more primitive system for visual telegraphy had been suggested in 1684 by Robert Hooke, but had not been tried out in practice. The semaphore was first employed in France, where the need for faster communication was felt after the revolution of 1789, when France was threatened by the allied forces of Britain, the Dutch Republic, Prussia, Austria and Spain. In 1794 the French engineer Claude Chappe completed an optical telegraph connection between Paris and Lille (230 km). It consisted of 22 semaphores over which a message could be transmitted with less than two minutes delay. Signaling rates were low; in the order of 10 to 20 signals per minute. Telegraphy networks were further extended in the following decades. In France the network ultimately covered a distance of 4800 km with 566 stations.

In the Netherlands the development of telegraphy also fell into the category "military research". After the Belgian revolt in 1830 a fast communication link was needed between the government in the Hague and the army in the southern part of the country, whose task it was to suppress the revolt.

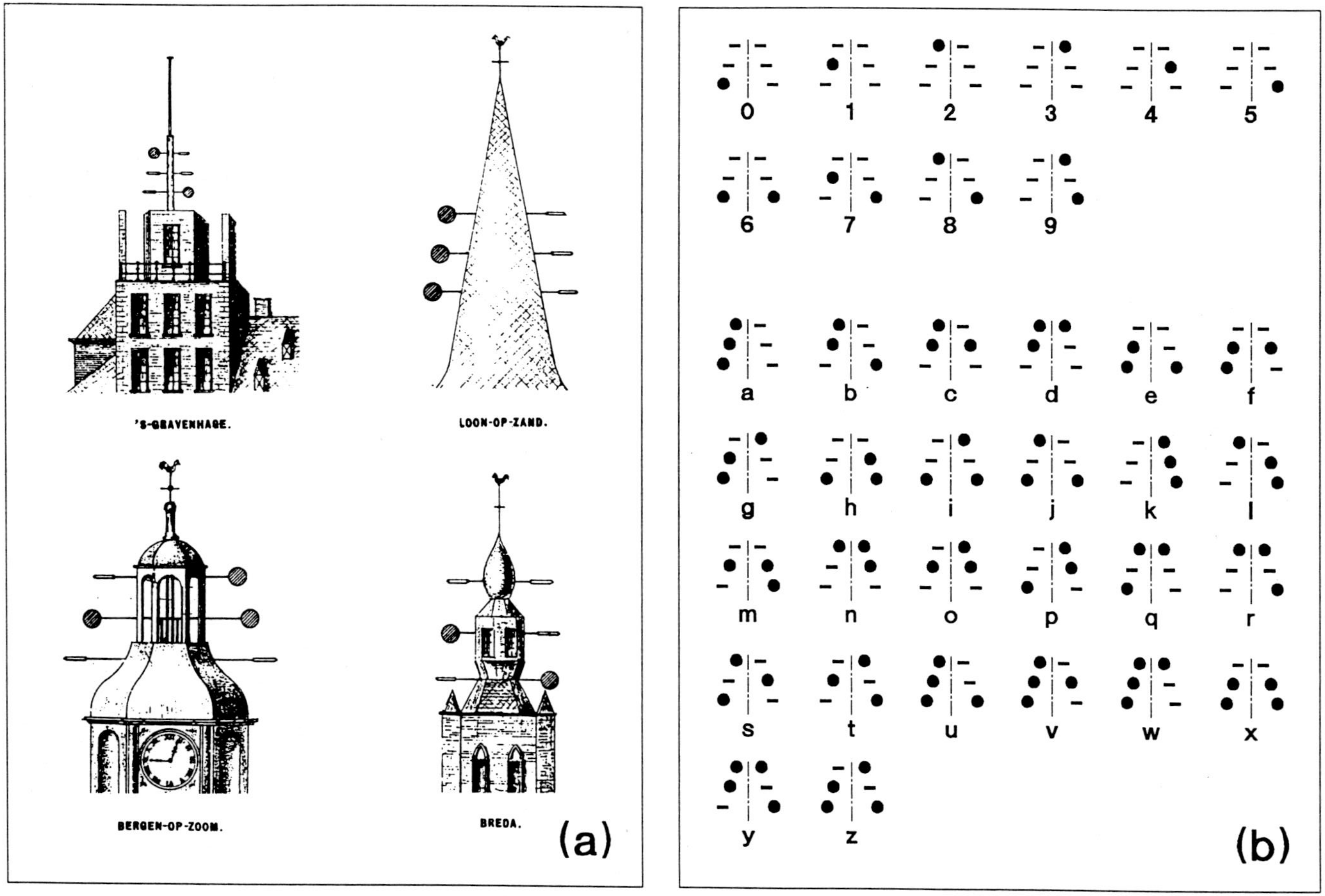

Figure 1.2 *(a) The semaphore and (b) the code developed by Antoine Lipkens in 1830/1831. (Source: W. Ringnalda, De Rijkstelegraaf in Nederland (The public telegraph in the Netherlands), courtesy: project group "History of Electrical Engineering", Delft University of Technology)*

Figure 1.2 shows the semaphore and the coding system developed by Lipkens, a Dutch telecommunication pioneer. Using this system the time required for transmitting a signal between the Hague and Flushing (19 stations) was reduced to less than 1 minute.

It is interesting to note that the development of modern telecommunication began with optical communication. By the middle of the 19th century optical telegraphy had developed as far as it could, however, and the torch was taken over by electrical telegraphy, which is less dependent on weather conditions and allows for higher signaling rates. Experiments with electrical telegraphy started already in 1809, when the German doctor Sömmering employed the gas-bubbles generated by an electric current in an electrolytic bath as a signal indicator. In 1845 Morse established the first intercity telegraph connection (Washington-Baltimore). His system was based on an electromagnet which was driven by the signaling current allowing for a higher signaling rate than that permitted by optical telegraphy or earlier experiments with electrical telegraphy.

Since then a variety of telecommunication services have emerged at an increasingly rapid rate [2]: speech transmission by telephone (1877, Bell), intercontinental radio telegraphy (1901, Marconi), radio broadcasting (1920), television broadcasting and telex (1930), mobile telephony, data-transmission, facsimile and color television (1960). More recently, services like broadband data transmission, telemetry, teletex, videotex, electronic mail and videoconferencing have become available. Figure 1.3 provides a detailed overview of the divergence since 1847, and the expected service range at the end of this century.

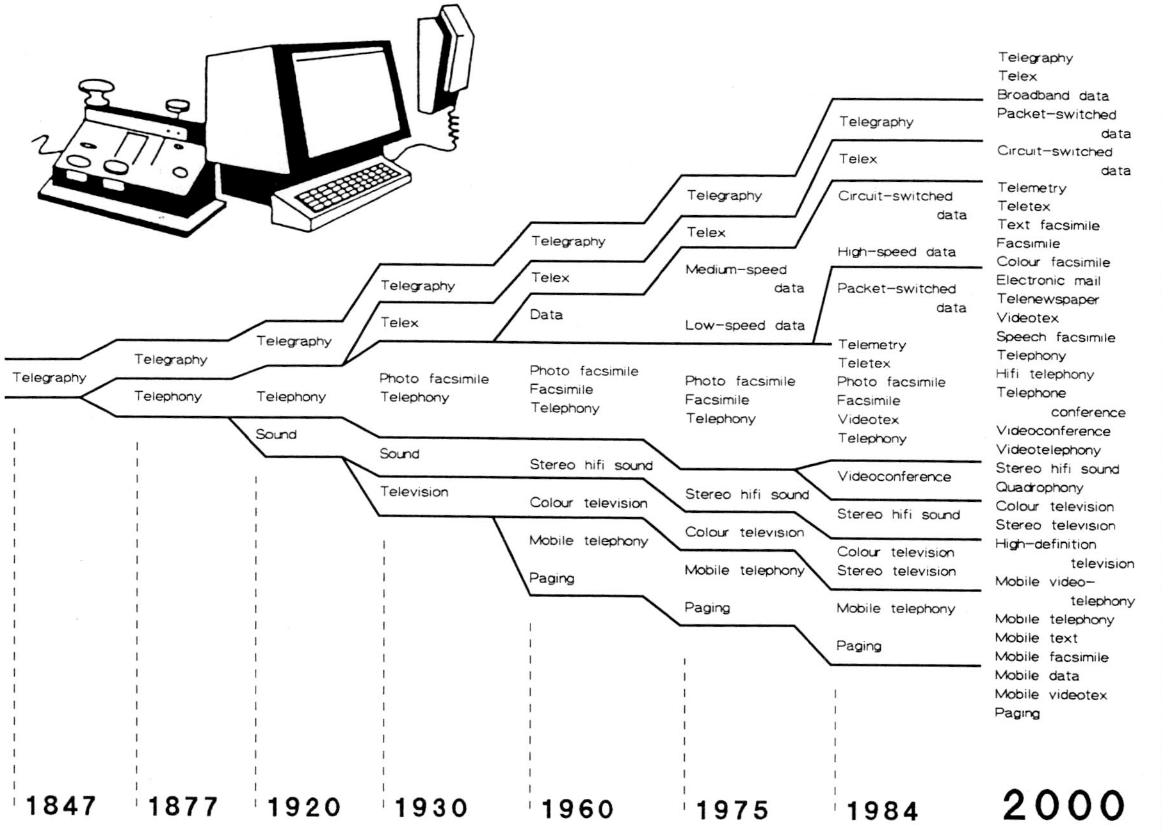

Source: Consortium British Telconsult/Consultel/Detecton/Nepostel/Sofrecom et al.

Figure 1.3 The development of telecommunication services from 1847 to 2000.

With respect to the networks we observe an opposite development; from many separate networks towards a few integrated ones. Radio and television distribution already use the same cable networks and are expected to be integrated with telephony and other services into an integrated subscriber network. The multitude of data communication networks which have appeared in the past decade are increasingly coupled with each other and with the public telephone network. These developments are heading towards a worldwide integrated communication network in the next century. Figure 1.4 shows an artist's impression of this network, as it may have evolved by the year 2010. An explanation is given below.

The private subscriber. In 2010 the private subscriber is still an important customer of the telecommunication companies. In addition to his telephone set, which is equipped with a video display, he has a terminal which gives him access to services such as electronic banking, teleshopping and many other low-priced information services. Next to his terminal he has a small color printer which is suitable for making *hard copies* and also serves as a facsimile printer. His old-fashioned television set has been replaced by a flat screen HDTV (High-Definition Television) set which is also suitable for inspection of electronic mail. All equipment is provided with an optical plug which is combined with the power supply plug and can be used everywhere in the house. The thermostat and the central heating apparatus communicate via the local inhouse network; other equipment such as the oven and an electronic door lock can be remotely controlled with a multi-purpose wireless controller. Bandwidth problems have disappeared; the capacity of the network is more than sufficient for the actual service package.

The professional user. His share in the use of the network has increased considerably since the introduction of the Integrated Services Digital Network. In addition to voice telephony, transfer of many kinds of data from simple electronic mail to documents with high-quality color graphics,

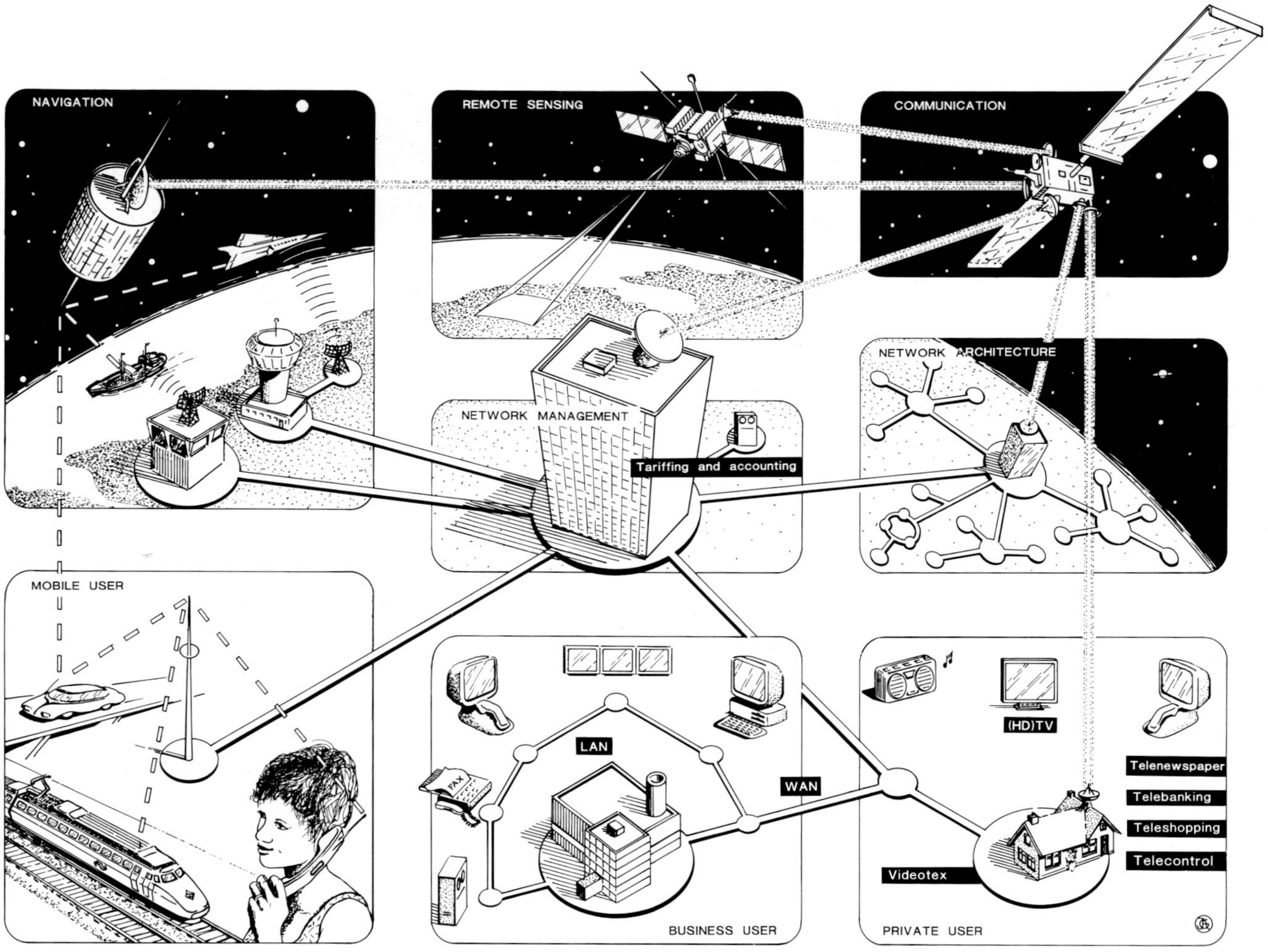

Figure 1.4 *An artist's impression of the world-wide integrated-services broadband communication network as it may evolve in the next century (courtesy: J.A. Geijp, Delft University of Technology).*

has become the most important application. Video-conferencing has also gained some popularity. The (single) inhouse network is intimately coupled with the public telephone network so that communication inside and outside the building is equally easy. An important difference with earlier networks is the worldwide standardization of communication protocols. All sorts of equipment communicate without difficulty over the network and wrong installation of equipment is reserved only to very intelligent people.

The mobile user. Car-drivers have taken their share in the spread of telecommunication services. The car telephone, which became popular in the 1980's, is now standard equipment in every car. Digital transmission and a sophisticated cellular structure of the mobile network guarantee fast access to the network even if traffic stagnation occurs and many drivers call at the same time. Personal portable telephone sets and mini-terminals, which can be used over a large part of the world, have become wide-spread.

Navigation. Automobiles as well as ships and airplanes have been equipped with an electronic navigation system. Using this system the on-board computer faultlessly guides the driver to the required destination, and avoids roadblocks and other stagnations, which are reported to the board computer by a mobile datalink. A traffic management system records the positions of trucks and guides them so as to avoid stagnations, thus reducing fuel consumption, air pollution and driver annoyance.

Remote Sensing. The quality of the weather forecast has been greatly improved on the basis of data provided by a large number of remote sensing satellites. Satellite data is transmitted to the network via a geostationary communication satellite. Ship movements, sea surface pollution, crop development and the state of the forests are also observed by satellite. Image data is processed in special centers and directly available to the interested (and paying) user via the network.

Satellite communication. The role of satellite communication has shifted from fixed communication links to more specific applications such as broadcast service to mobile users and thinly populated regions and communication with ships and aeroplanes. Furthermore geostationary satellites connect low-orbit satellites to the communication network via free-space laser beams, and provide a diversion path if a terrestrial connection is broken.

Network architecture and management. Behind the wall sockets of the network an invisible revolution has taken place. Each subscriber can attach as many terminals to his network port as he desires, without the need to ask permission and wait until the change has been effected. The local networks of different buildings can be coupled via the public network in such a way that two colleagues 1000 km apart communicate just as easily as if they were working in the same building. It is not necessary to reserve a hired line for this; the connection is realized within a few milliseconds when it is needed. The only occasion on which the user sees the difference between this type of connection and a permanent one is when he inspects his monthly bill!

The development outlined above is speculative, of course. Yet it is less of a question whether it will take place, than when.

From the foregoing the following trends can be discerned. Until the second World War, the development of electrical engineering was driven by telecommunication needs. Electron tubes, the predecessors of modern transistors, were initially developed for radio transmission. The close relationship between electronic and radio engineering in this period is still testified to by the name of the Dutch society of electrical engineers: Dutch Society for Electronics and Radio Engineering (NERG). Starting in the sixties the development became technology-driven, as a result of the incredible achievements in the field of micro-electronics, which shifted the attention of electronic engineers to computers and automation. Figure 1.1 shows that in the 1980's telecommunication has made a come back. Achievements in the field of computer technique and automation are presently being integrated into a telecommunication environment by the rapid spread of data networks

(LAN's, MAN's, and WAN's) [1] and such techniques as distributed processing.

It may therefore be expected that in the coming decades micro-electronic technology will become more and more telecommunication driven. This is already seen in III-V semiconductor technology, where the shift from AlGaAs to InGaAsP is fully attributable to the fact that optical fibers perform best at 1.3 and 1.55 µm wavelength. It is also seen in the increasing demand for high-frequency silicon processes, such as the BICMOS process, which combine complex digital functions with the high bit rates required in broadband communication networks and equipment. Telecommunication, which was at the birthplace of electrical engineering, is thus seen to regain its position after a short intermezzo.

A similar development is seen in optical communication technology. In order to integrate more and more services into the network function, the channels have to be as transparent as possible and the form in which data is transmitted has to be as universal as possible. These requirements have stimulated the development of broadband digital channels and switches. Important steps on the road to a completely transparent digital broadband network are the introduction of the Integrated Services Digital Network (ISDN), the development of an Integrated Broadband Communication Network (IBCN or Broadband (B)-ISDN), and the development of Asynchronous Transfer Mode (ATM) switching and routing techniques.

---

1) Local Area Networks, Metropolitan Area Networks, Wide Area Networks.

Optical communication links offer the bandwidth which is required for this development. It is therefore to be expected that the fixed part of the network will ultimately become all-optical. Short inhouse networks may be cabled with multimode fibers which provide easy connections combined with Gigabit transmission capacity on short distances. The other networks will be, or are already, cabled with monomode fibers which provide a virtually unlimited transmission capacity. Electronic switches will become bottlenecks in the transmission speed of the network. With optical switches Gigabit data streams are switched as easily as electronic mail. The development of small switching matrices and optical connectors which are suitable for mass-applications still has a long way to go, however, so that the switches will probably be the last parts of the network to become optical.

1.2 Optical communication technology

After the development of electrical telegraphy interest in optical signal transmission faded, but did not completely disappear. In 1880, for example, Bell experimented with a photophone based on photo-sensitive selenium cells in which speech was transmitted by means of a modulated light beam. The development of the first laser by Maiman [3] in 1960 (the phenomenon of stimulated emission had already been predicted by Einstein in 1917) gave a new impetus to optical communication research. Initially experiments were carried out with free space beam propagation and suffered from the same problem which hampered early optical communication, namely, great susceptibility to weather conditions.

In 1966, Kao and Hockam [4] proposed the use of optical fibers as a waveguide to eliminate environmental influences. The waveguiding properties of dielectric rods were already known in the nineteenth century and had been analyzed in 1910 by Hondros and Debye [5]. The main problem in employing glass fibers was the attenuation. Even the purest glasses exhibited attenuations in excess of 1000 dB/km, so that unamplified propagation was limited to some tens of meters. Kao and Hockam predicted that attenuations could be reduced below 20 dB/km. Their prophecy was already fulfilled in 1970 when Kapron, Keck, and Maurer [6] reported a (monomode) quartz fiber with 20 dB loss per kilometer. Within fifteen years fiber losses had been reduced below 0.2 dB/km, close to the theoretical limit.

Reproducible monomode fiber production and connection appeared to be too complicated for early fiber technology. Experiments with multimode fibers, driven with AlGaAs semiconductor lasers and LED's yielded more promising results. Multimode fibers reduced the tolerance requirements for fiber and connector fabrication technology, and the fastest route to a worldwide monomode fiber network appeared to be along the detour of multimode fibers.

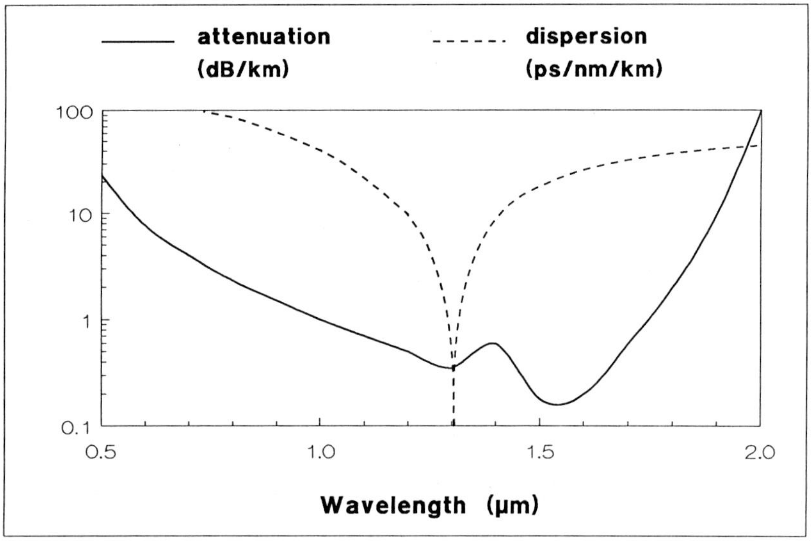

Figure 1.5 Optical attenuation and pulse dispersion of modern silica fibers.

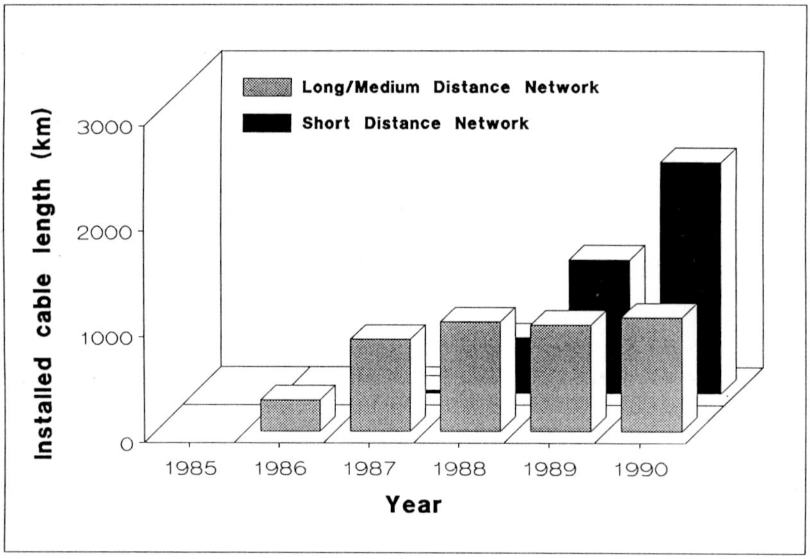

Figure 1.6 Yearly installed length of monomode optical fiber cables (containing 6 or 12 fibers) in the Netherlands (Source: PTT Telecom).

AlGaAs lasers provided a small (chip size: 0.3x0.3 mm^2) and relatively inexpensive alternative to gas and solid-state lasers, they could also be directly current modulated at bit rates of hundreds of Megabits per second, an important advantage. Until the mid-1980's optical communication was based on multimode fibers. Due to the non-linearity and noise problems in channels containing lasers in combination with multi-mode fibers, optical communication became almost completely digital.

The rapidly growing market for multimode fiber links formed a base for increased investments in fiber fabrication and connector technology. By the end of the seventies, technology had been developed so far that monomode fiber links, which offer a much greater transmission capacity, became feasible. In monomode fibers the transmission performance is limited by the fundamental material absorption and dispersion properties of quartz, which are shown in figure 1.5. It can be seen that the lowest attenuation occurs at 1.55 μm wavelength, whereas minimal pulse dispersion occurs at 1.3 μm. These wavelengths can not be generated by the AlGaAs semiconductor laser which are limited to the wavelength range of 0.78-0.87 μm. In the first half of the 1980's optical communication research shifted, therefore, to the quaternary InGaAsP semiconductor system on InP substrate. Employing this system the wavelength range from 0.9-1.65 μm, which includes the optimal wavelengths 1.3 and 1.55 μm, can be encompassed by an appropriate choice of the material composition.

Starting in 1984 monomode fiber technology outstripped multimode fibers at an unexpectedly rapid rate. Figure 1.6 shows the installed length of monomode fiber cables in the Netherlands in the years 1985-1989. The installation rate in the longer trunks is expected to drop soon; in the United States it has already dropped because the old network has almost been replaced. Monomode fiber is now seen to rapidly penetrate the short-distance network to the local exchanges. The main reason for this development is the increasing demand for bandwidth on communication trunks. For 140 Mb/s communication links, repeater spacings in coaxial waveguides are reduced to 2 km, and for 560 Mb/s to 1 km. Monomode fibers can transport even higher bitrates over more than hundred kilometers without amplification, so that transmission between cities is feasible without repeaters in most parts of Europe. Already more than half of the world's intercity traffic is carried by monomode fiber links [7].

The following step towards a completely integrated digital communication network will be the extension of the fiber network to the home, this will increase the optical communication market by at least one order of magnitude. As we have seen before, however, history does not always take the straight line from one point to another. The road from present digital optical transmission to an integrated digital network may very well go along the path of analog optical transmission [8]. The reason for this is the inability of the existing coaxial cable networks to transport the bandwidths required by High-Definition Television, which is expected to appear on the market around 1995. Cable-television network operators will, therefore, have to adapt their networks in the short term, and monomode fibers are considered to be the most economical alternative for carrying the signals to the subscribers.

If the digital transmission scheme of modern optical links is applied in the new distribution network, all subscribers will have to be equipped with fast digital decoders. This requires enormous investments over a short period. To avoid this problem research is increasingly focused on Subcarrier Modulation (SCM), an analogue transmission technique in which a composite electrical signal, containing the full desired spectrum of video and FM-channels, is intensity-modulated onto an optical carrier. After detection this signal can be directly applied to conventional receivers. The quality of modern lasers, in combination with monomode fibers, meets the linearity and signal-noise requirements imposed by this modulation scheme. If component costs can be sufficiently reduced, fiber-to-the-home may become reality for many subscribers before the end of the century.

A big problem on the road to technical integration of the cable-television network and the public telephone network is formed by the ownership and the exploitation of the networks. In many countries the cable-television networks and the public telephone network are operated by different companies. A rapid introduction of an integrated subscriber network is only possible if agreement is achieved on a common exploitation of the network infrastructure.

1.3 Integrated optics

The term "Integrated Optics" was introduced in 1969 by Miller [9], in an issue of the Bell Systems Technical Journal containing several contributions on planar optical waveguides. In his article "Integrated optics: an introduction" he discussed the potential of planar technology for fabrication and integration of optical components and circuits. The first decade (1970-1980) of "Integrated optics" was actually "Planar optics": research was focused on single components; integration of several components was technologically out of reach. The laser was the only planar component which passed beyond the R&D-stage.

The development of integrated-optics technology was hampered by the fact that optical communication was initially based on multimode fibers. Planar technology is not particularly suited to the fabrication of multi-mode devices. Further, in coupling light from a multi-mode fiber to a monomode device most of the power is lost. Research on components other than lasers was, therefore, conducted at a low profile, and restricted to passive dielectric materials like oxides and compound glasses.

In the second decade (1980-1990) laser technology was the first planar-optics activity to become *big business*, mainly because of its application in compact-disc players. In this decade, the break-through in the area of monomode fibers formed the base for a telecommunication-driven acceleration of integrated-optics research. In addition to the development of sophisticated long-wavelength telecommunication lasers, a variety of planar components and a few integrated circuits (coupling and switching matrices) were reported.

The third decade (1990-2000) is destined to become the decade of really integrated optics. In the last few years, activities on integration of photodetectors with field-effect transistors have been reported at each conference. Recently an integrated coherent receiver and a multiplexed multi-wavelength transmitter with three lasers which can be modulated independently were reported. The main driving force behind integrated-optics research is its potential to avoid most of the connection costs which occur in optical circuits built from fiber-pigtailed components, and to reduce the large volume occupied by these circuits [10].

Figure 1.7 shows an artist's impression of a coherent integrated optical receiver, such as may be installed in large numbers in the homes of future network subscribers. The operation is identical to that of a heterodyne radio receiver: the signal received from the fiber is mixed with that of a local oscillator in order to realize an IF signal at the frequency difference between the mixed signals. Mixing occurs in the photodiodes shown in the figure after the signals have been combined in a 2x2-port directional coupler. A very stable laser acts as a local oscillator (with a frequency around 200,000 GHz); its frequency is controlled by a feedback loop designed to maintain the IF frequency at a stable value of a few Gigaherz.

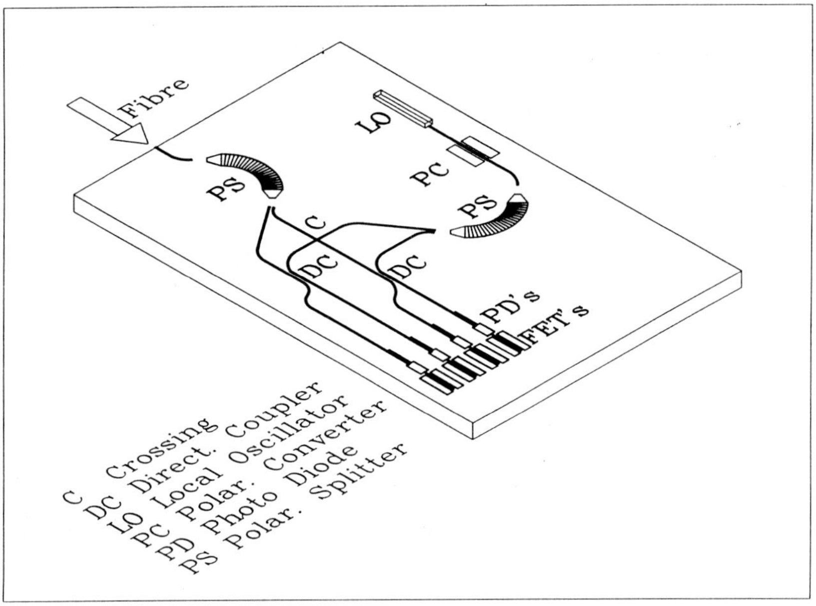

Figure 1.7 *An artist's impression of a monolithically integrated coherent optical receiver (courtesy Y.S. Oei, Delft University of Technology).*

Effective mixing only occurs if the combined signals have the same polarization state. If the polarizations of the signals are orthogonal, no mixing products are generated; the receiver will experience total fading of the signal. To avoid this the received signal, which has an undefined and slowly fluctuating polarization state, is split into two orthogonal polarizations by means of a polarization splitter. The same is done with the local-oscillator signal, which must first be rotated over 45° by a polarization controller in order to produce equal power in both polarizations. Both polarizations are processed in parallel and added electronically after detection. The sum signal is no longer dependent on the polarization state of the received signal. This technique is known as polarization diversity.

Monolithic realization of a chip as discussed above, which contains sources and detectors, is only possible in III-V semiconductor materials, notably the quaternary alloy InGaAsP (or InGaAlAs) on InP-substrate, because it is the only material system which allows for generation and detection of light at the wavelengths 1.3 and 1.5 µm where the fiber properties are optimal. The InP-based quaternary semiconductors are, therefore, the most promising materials for monolithical integrated circuits.

In addition to the quaternary semiconductors, a few other materials will also play a role in future optical networks. An important factor regarding the application of an integrated optical circuit is the waveguide type. Roughly speaking, two waveguide types can be distinguished: low-contrast and high-contrast waveguides. An optical fiber with a waveguide size between 5 and 10 µm and an index contrast below 0.01 is a typical low-contrast waveguide. Quaternary semiconductor lasers contain high contrast waveguides; effective waveguide dimensions are in the order of 1 µm or smaller, index contrasts in the order of 0.1.

Figure 1.8a and 1.8b depict the waveguide geometries of the fiber and the laser in realistically scaled proportions. For easy coupling to sources, especially lasers, waveguide dimensions and contrasts should match those of the laser, which is of the high-contrast type. High-contrast waveguides (figure 1.8b) allow for small bending radii and, consequently, a small device area. The dimensions of tapers, Y-junctions and couplers also scale down with increasing index contrast. A disadvantage of high-contrast waveguides are the

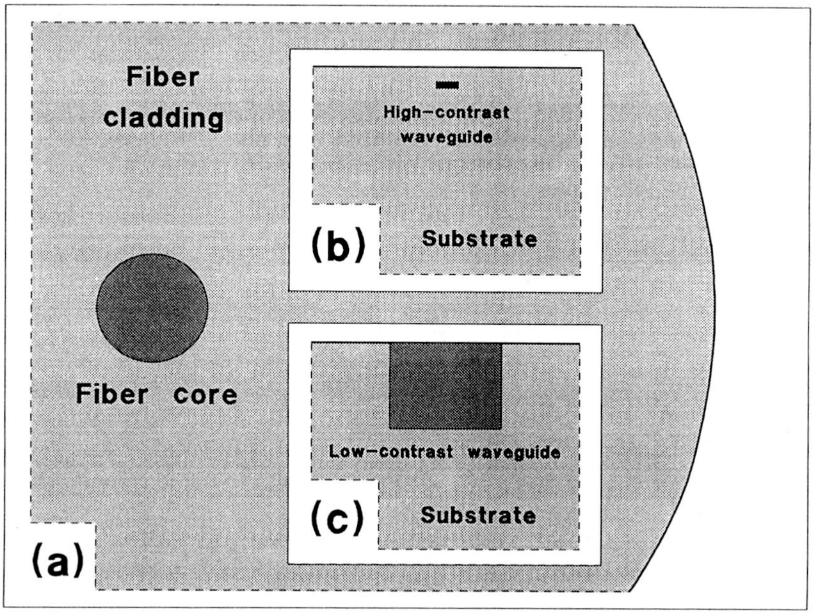

Figure 1.8 Different waveguide types in realistically scaled proportions:
(a) monomode fiber,
(b) laser and high-contrast waveguide,
(c) low-contrast waveguide.

complications in coupling to fibers. For easy coupling to fibers, a low index contrast and dimensions comparable to the fiber core are desirable (figure 1.8c). However, this advantage is off-set by a larger circuit size.

Integrated circuits containing lasers are necessarily of the high-index type. For other circuits, the cost balance between low- and high-index chips is dominated by the low-index chip cost (circuit area) and the high-index fiber-coupling costs. With increasing circuit complexity the balance will shift towards high-index waveguides, especially for chips which are required in large numbers in the local network. High-index chips are, therefore, expected to become the dominant waveguide type for application at the network terminals. Low-contrast chips, in passive (silicon-based TEOS glass) or electro-optical materials (lithium niobate or electro-optical polymers), will continue to play a role in the network nodes.

The development of monolithically integrated optical chips, which is presently growing out of its infancy, will probably take another decade to reach the low-cost commercial production stage. The introduction of fibers in the local network will require a similar period. Large-scale application of optical chips may, therefore, begin around the turn of the century.

1.4 Integrated-optics research at the Delft University of Technology

Delft research on integrated optics was began in 1982. A decision was taken to gain experience in this field with aluminum oxide as a waveguide material, and to switch to the more challenging, but also more complicated, III-V semiconductors at a later stage. Sputtering experiments began in 1983; it took almost a year to develop a sputtering and annealing process which yielded reproducible low-loss aluminum oxide films. In the following years, a technological infrastructure for fabrication and characterization of integrated optical components was developed. The most important activities in the period covering 1984 and 1985 concerned the development of equipment for accurate measurement of waveguide properties (effective index, propagation loss and mode profile) and transmission properties of planar components, and further a lithographic process for dry etching 2-μm waveguides, and software extensions to existing layout programs for accurate generation of waveguide patterns including bends.

Parallel to the development of fabrication technology and characterization, component research commenced with the development of low-loss straight-waveguides [11,12], followed by a research project on waveguide bends. The latter project provided the Delft University with a leading position in the field of low-loss short waveguide bends [13,14], a position which it continues to occupy. Miniaturization and insensitivity to process tolerances have become leading themes in selecting research subjects. Results have been achieved in the field of small-size planar tapers [15], planar lenses [16,17] (in cooperation with Twente University), polarization splitters [18,19], wavelength demultiplexers [20,21], and couplers [22,23].

In 1984 the Ministry of Economic Affairs granted funds for starting a research project on Integrated Optics in InGaAsP, in cooperation with the Optics Research Group of the department of Applied Physics and the Dr Neher Laboratory of the Dutch PTT, with technological support of Philips Research Laboratories. The project started in 1985 with the development of a liquid-phase epitaxy reactor, which became operational by the end of 1986. In the following years processes were developed for growing double-hetero waveguide structures, wet etching and Reactive Ion Etching of waveguides, and characterization of components. Research is aimed at transfering component performance obtained in the aluminum oxide system to the InGaAsP system. In 1989 the universitie's leading position with respect to extremely short waveguide bends was extended to InGaAsP waveguides, in cooperation with Philips Research Laboratories [24]. Present research includes the realization of directional couplers, coupling networks and polarization splitters. In 1989 a research project involving the integration of passive waveguide circuits with detectors and field-effect transistors into an optical receiver chip was begun.

The above program, which is conducted in cooperation with the Optics Research Group of the Delft University, the Neher Laboratory of the Dutch PTT, Philips Research Laboratories, and the Universities of Twente, Eindhoven, and Ghent (Belgium), has provided the Delft University with a solid base for participation in the international development of Integrated Optics.

1.5 The subject of this thesis

The present thesis covers the work on the (silicon-based) aluminum oxide waveguide system with which integrated-optics research started in Delft. It handles technological research (chapter 2), device modeling and experiments (chapters 3 to 6), measurement technique (chapters 7 and 8), and fiber coupling (chapter 9). Parts of this work have been published, references are included in the summary:

Chapter 2 describes the development of a process for fabrication of low-loss waveguiding films for application both at short (633 and 780 nm) and long wavelengths (1300 and 1550 nm) [11,12].

Chapter 3 describes the waveguide structure used in the experiments.

Chapter 4 builds on the work of Pennings [25] on waveguide bends, and describes an extension to his normalized design strategy.

Chapter 5 describes design, modeling and experiments on short directional couplers with reduced susceptibility to production tolerances [22,23].

Chapter 6 describes design, modeling and experiments on optical phased arrays, applied as polarization splitters and wavelength demultiplexers [16,18-21]

Chapter 7 describes the results of an investigation into the measurement accuracy of ellipsometry which is used for measuring refractive indices of planar films [26-28].

Chapter 8 describes the prism-coupling technique which was developed for accurate characterization of films, waveguides and devices [15,29].

Chapter 9 describes a novel method for low-loss coupling of high-contrast optical waveguides to monomode fibers [30].

Although the aluminum oxide waveguide system is of limited interest for commercial applications (because of its high index contrast) it is very attractive for component research because of the good reproducibility of the fabrication process (better than 10^{-3} for the effective index), and because of its easy and reproducible optical accessibility for characterization of devices and circuits. An additional advantage is the fact that the effective wavelength of visible He-Ne light inside the waveguide material (0.4 µm) equals that of 1.3 µm radiation in InGaAsP waveguides. Furthermore, the index contrasts are comparable so that aluminum oxide may be used for prototyping InGaAsP components with visible light. As such, it has proven to be of great value.

References

1.1 Telecommunications for Europe 1992, The CEC Sources, IOS, Amsterdam, Springfield, VA., Tokyo, 1989.

1.2 From Semaphore to Satellite, publication of the International Telecommunication Union, Geneva, 1965.

1.3 T.H. Maiman, "Stimulated optical radiation in ruby", Nature, Vol. 187, pp. 493-494, 1960.

1.4 K.C. Kao and G.H. Hockam, "Dielectric fibre surface waveguides for optical frequencies", IEE Proceedings, Vol. 113, pp. 1151-1158, 1966.

1.5 D. Hondros and P. Debye, "Elektromagnetische Wellen an die elektrischen Drahten", Ann. Phys., Vol. 32, pp. 465-476, 1910.

1.6 F.P. Kapron, D.B. Keck, and R.D. Maurer, "Radiation losses in glass optical waveguides", Applied Physics Letters, Vol. 17, pp. 423-425, 1970.

1.7 P.C. Cochrane, "Future directions in long haul fibre optic systems", British Telecom Technology Journal, Vol. 8, No. 2, pp. 5-17, 1990.

1.8 S.S. Wagner and R.C. Menendez, "Evolutionary architectures and techniques for video distribution on fibers", IEEE Comm. Magazine, Vol. 1, no. 4, pp. 17-25, December 1989.

1.9 S.E. Miller, "Integrated Optics: An introduction", Bell Systems Technical Journal, Vol. 48, pp. 2059-2069, 1969.

1.10 T.L. Koch and U. Koren, "Photonic Integrated Circuits: Research curiosity or packaging common sense?", IEEE Lightwave Communication Systems Magazine, Vol. 1, No. 4, November 1990, pp. 50-56.

1.11 M.K. Smit, G.A. Acket and C.J. van der Laan, "Al_2O_3 films for integrated optics", Thin Solid Films, Electronics and Optics, Vol. 138, pp. 171-181, 1986.

1.12 M.K. Smit, "Sputtered planar optical waveguides", Sensors & Actuators, Procs. S&A Symposium, Enschede, The Netherlands, October 30-31, 1986, pp. 45-55, Deventer: Kluwer Technical Books, 1986.

1.13 E.C.M. Pennings, G.H. Manhoudt and M.K. Smit, "Low-loss bends in planar optical ridge waveguides", Electronics Letters, Vol. 24, pp. 998-999, 1988.

1.14 E.C.M. Pennings, J. van Schoonhoven, J.W.M. van Uffelen, and M.K. Smit, "Reduced bending and scattering losses in a optical 'double-ridge' waveguide", Electronics Letters, Vol. 25, pp. 746-748, 1989.

1.15 M.J.M. van Vaalen and M.K. Smit, "Fabrication and measurement of low-loss planar optical tapers", Fiber and Integrated Optics, Vol. 7, pp. 235-239, 1988.

1.16 M.K. Smit, "New focusing and dispersive planar component based on an optical phased array", Electronics Letters, Vol. 24, pp. 385-386, 1988.

1.17 P.C.A. Hammes, G.J.M. Krijnen, A. Driessen, and M.K. Smit, "High efficiency small-size planar hyperbolical lenses", Electronics Letters, Vol. 26, pp. 1064-1065, 1990.

1.18 A.R. Vellekoop and M.K. Smit, "Low-loss planar optical polarisation splitter with small dimensions", Electronics Letters, Vol.25, pp.946-947, 1989.

1.19 A.R. Vellekoop and M.K. Smit, "A small size polarisation splitter based on a planar optical phased array", Journal of Lightwave Technology, Vol. 8, pp. 118-124, 1990.

1.20 A.R. Vellekoop and M.K. Smit, "A polarization independent planar wavelength demultiplexer with small dimensions", Procs. Eur. Conf. on Opt. Integr. Systems at Amsterdam, The Netherlands, September 25-28, 1989, paper D3.

1.21 A.R. Vellekoop and M.K. Smit, "A polarisation-independent four-channel wavelength demultiplexer based on a planar optical phased array", Journal of Ligthwave Technology, Vol. 9, pp. 310-314, 1991.

1.22 L.B. Soldano, A.H. Dubost, F.B. Veerman, M.K. Smit, B.H. Verbeek, and E.C.M. Pennings, "Multimode Interference Couplers", Technical Digest of the Integrated Photonics Research Topical Meeting, Monterey, California, USA, April 9-11, 1991.

1.23 F.B. Veerman, "An optical passive 3-dB TMI-coupler with reduced fabrication tolerance sensitivity", to appear in Journal of Lightwave Technology.

1.24 B.H. Verbeek, E.C.M. Pennings, J.W.M. van Uffelen, and P.J.A. Thijs, "Fabrication and analysis of low-loss InGaAsP/InP optical waveguides with extremely small bends", Procs. 15th Eur. Conf. on Opt. Comm. (ECOC'89), September 10-14, Gothenborg, Sweden, Volume 3: Post-deadline papers, PDB-9, pp. 78-81, 1989.

1.25 E.C.M. Pennings, Bends in optical ridge waveguides, Modeling and Experiments, Ph.D. Thesis, Delft University of Technology, 1990.

1.26 J.W. Verhoof and M.K. Smit, "Error analysis for ellipsometry", Procs. 12th IMACS Congress, Paris, July 18-22, Vol. 2, pp. 502-504, 1988.

1.27 J.W. Verhoof and M.K. Smit, "Accuracy analysis for multiple angle of incidence ellipsometry", Thin Solid Films, Electronics and Optics, Vol. 189, pp. 193-203, 1990.

1.28 J.W. Verhoof and M.K. Smit, "A bounded-error approach to accuracy analysis in ellipsometry", Mathematics and Computers in Simulation, Vol. 32, pp. 545-551, 1991.

1.29 W.A. Pasmooij, M.K. Smit and P.A. Mandersloot, "Prism-coupling of light into narrow planar optical waveguides", Journal of Lightwave Technology, Vol. 7, pp. 175-180 (1989).

1.30 M.K. Smit and A.H. de Vreede, "A novel method for matching the mode profile of planar waveguides to single-mode fibers", to appear in Applied Optics.

Chapter 2

Aluminum oxide films for integrated optics

In this chapter the fabrication process and the properties of the aluminum oxide waveguide films are described. Special attention is paid to the heat treatment, which is applied to reduce optical film loss.

2.1 Introduction

The most important characteristics of the films to be used in integrated optical circuits are thickness, optical attenuation and refractive index, as well as the production tolerances of these attributes and the chemo-mechanical stability of the films.

Silicon wafers were selected as a substrate material after some initial experiments with Corning-7059 glass substrates. Silicon wafers provide the advantage of compatibility with conventional semiconductor technology and equipment, and have, in addition, the potential to integrate vast quantities of micro-electronics and also short-wavelength optical detectors with the optical circuit. Furthermore, they have excellent surface quality at a relatively low price and they withstand the high temperatures applied during heat treatment. Finally, the absorption of silicon for short wavelengths is advantageous for high-resolution optical pattern definition (no undefined reflections from the substrate bottom). If silicon is chosen as a substrate material, silicon dioxide is the natural choice for the cladding layers. As a waveguiding material a class of oxides, nitrides, fluorides, silicates and compound glasses is available. Many materials have been investigated for application in integrated optical circuits [1-11].

Easy and accurate optical measurements are possible for films with guided modes having effective indices of maximally 1.7 because of the availability of prisms and matching liquids with indices up to this value. This point will be discussed in more detail in chapter 8. For materials with considerably higher refractive indices such as silicon nitride, which gained some popularity because of its compatibility with silicon technology, the effective index can be kept sufficiently low by applying very thin films. In such films, however, orthogonally polarized modes are strongly degenerate and the difference in coupling lengths is considerable. Therefore, it is no longer possible to optimize the waveguide structure such that both polarizations can be excited in a reproducible way. This increases the difficulty of film- and device characterization.

Aluminum oxide is one of the few oxides having a refractive index in the intermediate range between silicon dioxide (n≃1.45) and silicon nitride (n≃2). This allows for well-controlled prism coupling without having to apply very thin films. Its refractive index varies between 1.55 and 1.75, dependent on deposition conditions. Aluminum oxide is widely applied in thin film technology. It is used as a protective and anti-reflection coating in solar cells and lasers, as semiconductor passivation layer, as a dielectric layer in thin-film capacitors, as diffusion mask, and as an insulating layer in multi-layer interconnection schemes, to mention a few examples. A recent development is the hetero-epitaxial growth of aluminum oxide on silicon substrates [12,13] which may render an important role to this material in three-dimensional integration.

Little has been published on the application of aluminum oxide for integrated optics. Until 1981, when we decided to begin research on rf-sputtered aluminum oxide films, only three publications mentioned the use of this material for integrated optics. Goell [4] reported that barium silicate films have good optical properties, but that aluminum silicate films are quite lossy. Kersten et al. [6] investigated the optical properties of a number of evaporation-deposited oxides and fluorides, among them aluminum oxide. They found no attenuation values lower than 10 dB/cm for any of their films, which raises some doubt about their preparation method. Kiselev and Red'ko [8] give a plot suggesting very low attenuation values for rf-sputtered aluminum oxide films, but provide no information on the values actually measured. In 1984 low-loss aluminum oxide films were reported by Binh et al. [10], using ion-beam-assisted evaporation, and Este and Westwood [11], using modified dc magnetron sputtering.

Delft research on rf-sputtered aluminum oxide waveguides (see figure 2.1) actually started in 1982. By early 1983, measurement equipment had been developed so far that reliable attenuation measurements could be performed. In that year a large number of experiments were carried out in order to reduce the optical losses which appeared to be persistently high (20 dB/cm or more). In the fall of 1983 it was discovered that film losses were reduced by heat treatment [23]. In a subsequent systematic investigation on the effect of heat treatment in combination with the process parameters, film losses were reduced below 1 dB/cm at a wavelength of 633 nm. The results of this investigation were published in 1986 [14,15]. More recently, attenuation as low as 0.35 dB/cm at 1300 nm wavelength has been achieved.

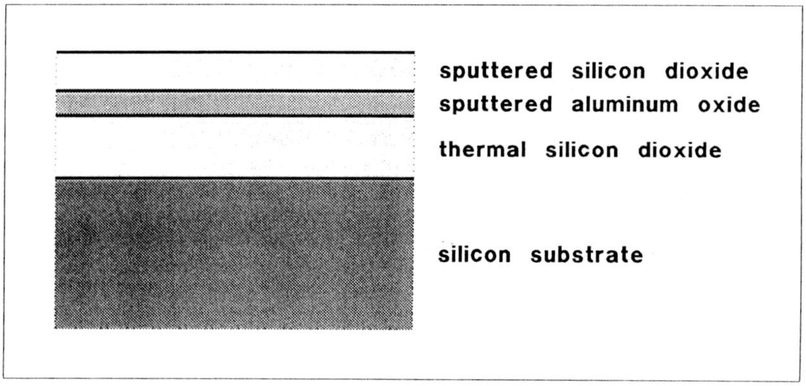

Figure 2.1 Transverse waveguide structure.

After the successful reduction of propagation loss, emphasis shifted towards device research. Some additional investigations into the structure of the films and the mechanism of the spectacular attenuation reduction have been carried out, however, and these will be reported in the sequel.

2.2 Waveguide fabrication

2.2.1 Substrate preparation

The high refractive index of silicon (n>3) requires the application of a transparent low-index buffer layer to isolate the aluminum oxide waveguide optically from the silicon substrate. For this purpose we use a silicon dioxide layer obtained by thermal oxidation of the wafer. For the waveguides applied (see chapter 3) an oxide thickness of 2 μm has proven to be sufficient for isolating the guided modes from the silicon substrate for wavelengths up to 820 nm. A 4 μm thick layer is sufficient for wavelengths up to 1550 nm.

If the silicon wafers are not processed immediately after oxidation, thorough cleaning is necessary in order to avoid increased scattering losses due to surface contamination. Excellent results are obtained by ultrasonic cleaning in a strongly alkaline soap solution (ROTH RBS-50, 10%) for two hours, rinsing for one hour in demineralized water to remove soap residues, and finally spin-drying the substrate. Surface quality is visually inspected by observing the scattering of the wafer surface on illumination by a He-Ne laser beam; well-cleaned substrates show very little scatter.

Immediately after cleaning the substrates are placed in the vacuum deposition chamber in which they are kept under vacuum for at least two hours. No sputter etch is applied prior to sputtering; film adherence is excellent without prior sputter etching.

2.2.2 Deposition process

A detailed description of the sputtering process is given by Chapman [16]. For our investigations we applied a cryo-pumped Alcatel SCM-600 sputtering system (figure 2.2) with 6" targets and a water-cooled substrate table, which was installed in the summer of 1983. Aluminum oxide films were sputtered with 99.999% argon gas from a 99.999% Al_2O_3 target with the rf diode

Figure 2.2 *The Alcatel SCM-600 sputtering system, in which the aluminum oxide films are deposited.*

process. This process yields the best thickness uniformity across the wafer. All circuits were covered with a silicon-dioxide layer to protect them against environmental influences. The thickness of this layer is less critical. It was, therefore, sputtered with the rf magnetron process which yields higher sputter rates but less uniform film thickness.

The effect of sputtering pressure on film quality has been investigated by several authors. Salama [17] reports a reduction of the refractive index with increasing pressure, indicating increased film porosity. The same effect was found by Westwood [18] and Paulson [19] for sputtered Ta_2O_5, and by Taguchi [9] for evaporated ZrO_2. In addition Paulson [19] found a positive correlation between optical loss and sputtering pressure. We therefore sputtered our films at a pressure of 0.8 Pa, which is the lowest pressure at which a stable plasma could be maintained.

Although sputtering is done from an oxide target, some reactive gas (O_2) has to be applied to the vacuum chamber in order to re-oxidize the small fraction of metallic aluminum which arises from decomposition at the target. Goranchev [20,21] found that even for a metallic aluminum target 2% oxygen at a total sputtering pressure of 10 Pa is sufficient to fully oxidize the target. Paulson [19] reports good quality films being sputtered with 10% oxygen at sputtering pressures from 0.5 to 1.5 Pa, and increasing optical loss for higher oxygen content. We obtained good results with an oxygen fraction of 10%.

Experiments were conducted with rf power densities between 2.5 and 5 W/cm^2 and a substrate bias varying from 0-150 V. The sputter rates for films sputtered without bias vary from 3.5 nm/min for 2.5 W/cm^2 to 7 nm/min for 5 W/cm^2. Sputter rates for bias-sputtered films vary between 6 and 9 nm/min without strong dependence on the rf power density; this may be explained by increased film porosity at lower sputter rates. Dependence of film properties on sputtering parameters is presented and discussed in section 2.3.

2.2.3 Annealing procedure

The optical attenuation of the films is invariably high on leaving the sputtering chamber. Angenent [23] found that the attenuation can be reduced by heat treatment. To study the effects of heat treatment a number of films, sputtered under different conditions, were given a series of 30-minute treatments at increasing temperatures. The optical attenuation was measured after each treatment (with a sliding prism as discussed in chapter 8). For all investigated films the obtained anneal curves had a qualitatively similar shape, showing a gradual reduction of the attenuation with increasing temperature up to a temperature T_{min}, and a dramatic increase accompanied by visually observable film damage beyond T_{min}. Figure 2.3 shows an experimentally determined anneal curve for the attenuation at 1300 nm wavelength of a 0.5 µm thick film.

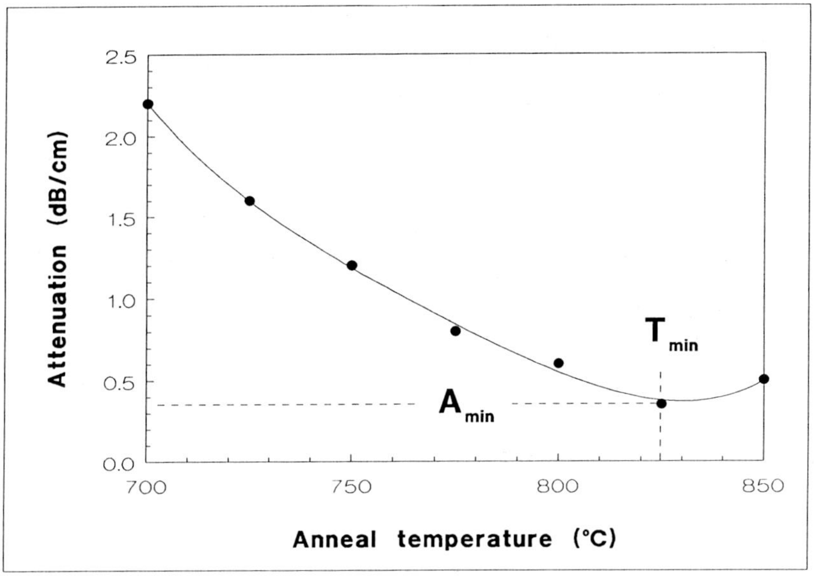

Figure 2.3 Anneal curve of a sputtered aluminum oxide film. Each point represents the optical attenuation measured at 1300 nm wavelength after a (cumulative) 30-minute anneal step. The curve was recorded for a 0.5-μm film sputtered with 5 W/cm² rf power without substrate bias.

The optimum of the curves as shown in figure 2.3 can be obtained in a single anneal step. Jongeling [24] found that the effects of annealing temperature and time on the attenuation are exchangeable: a 25 K temperature increase gives the same effect as doubling the annealing time. The cumulative effect of previous treatments, relative to the last one, thus follows as a series ½ + ¼ + ... ≃ 1. The same effect as a series of 30-minute treatments at temperatures increasing in steps of 25 K is, therefore, obtained with a single 60-minute treatment at the temperature of the last step.

For films sputtered with 5 W/cm² rf power and zero substrate bias (these appear to perform best as shown in the next section), we found annealing times of 55 minutes at 800 °C to be optimal for 0.25 µm thick films[1], and 60 minutes at 825° C for 0.5 µm films[1] to be optimal. These results are in good agreement with recorded 30-minute anneal curves.

2.3 Film properties

2.3.1 Optical attenuation

The lowest attenuation achieved was 0.35 dB/cm at 1300 nm wavelength, 0.5 dB/cm at 780 nm, and 1 dB/cm at 633 nm wavelength[1]. Figure 2.4 shows the experimentally established dependence of the minimal attenuation A_{min} and the corresponding anneal temperature T_{min} on the applied bias voltage. Stable substrate bias below 50 V being technically impossible, no results are shown for low bias voltages other than zero.

From this figure it is clear that films sputtered without bias yield the lowest optical attenuation and are, therefore, most interesting for use in integrated optical circuits. Experiments on films sputtered without bias and with varying rf power indicate that in the range from 2.5 to 5 W/cm² the rf power has little

1) An attenuation of 0.35 dB/cm @1300 nm was obtained for a 0.5 µm thick film, sputtered with 5 W/cm² rf power (900 W) and zero substrate bias, after annealing 60 minutes @825 °C. An attenuation of 1 dB/cm @633 nm was obtained for a 0.25 µm thick film, sputtered under the same conditions, after annealing 55 minutes @800 °C.

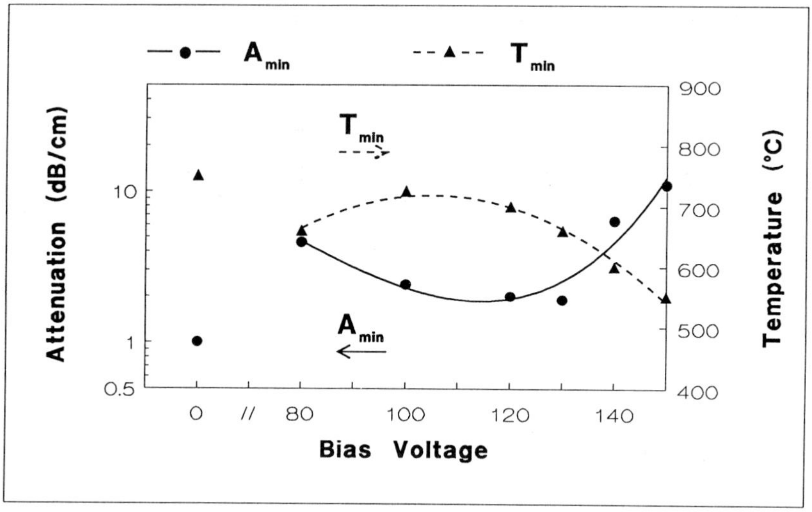

Figure 2.4 Dependence of the minimal attenuation (@ 633 nm) and the optimal anneal temperature on the applied substrate bias voltage, for 0.25-μm films sputtered with 5 W/cm² (900 W) rf power.

influence on the film properties. For bias-sputtered films, the film properties are strongly dependent on the applied bias voltage. The lowest attenuation obtained for bias-sputtered films (after annealing) was 2 dB/cm (@633 nm).

2.3.2 Refractive index

The refractive index of films sputtered without bias, as determined from the propagation constants of the fundamental modes (see chapter 8, subsection 8.4.2), is 1.69 (@633 nm, after annealing). Ellipsometric measurement yields an even higher index value of 1.70. The reason for the discrepancy between the two different measurement methods lies in the fact that the films are slightly birefringent (Δn 0.002), which is not accounted for by either method.

These experimental results are discussed in more detail in chapter 8, subsection 8.4.2. The ellipsometrically measured value of 1.70 is expected to be most representative; similar values have been determined for thick (multimode) films from the propagation constants of two modes having the same polarization. These values are the highest refractive index values reported in the literature for low-loss aluminum oxide films measured in the red spectral window.

For design purposes the lower index value as determined from the propagation constants of the two fundamental modes is most suitable. It permits the performance of computations as if the material were isotropic, thus avoiding the need for more complicated methods involving anisotropy. If not explicitly stated otherwise, values mentioned in the text were measured using the latter method.

In the range from medium (2.5 W/cm^2) to high power density (5 W/cm^2) the high index appears to be virtually independent of the applied rf power. For 1 W/cm^2 a slightly lower index value (1.67 @633 nm) was measured. The refractive index of bias sputtered films is substantially lower than that of zero-bias sputtered films, and strongly dependent on the applied rf power. Measurement results vary from as low as 1.56 for films sputtered with 2.5 W/cm^2 (without significant dependence on bias voltage) to 1.67 for films sputtered with 5 W/cm^2 (with a weak dependence on bias voltage). Figure 2.5 shows the experimental results.

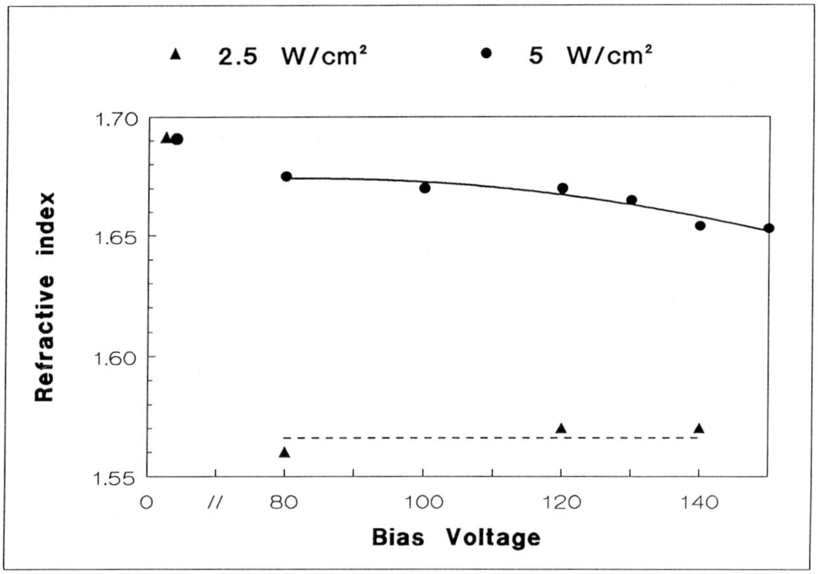

Figure 2.5 Dependence of the refractive index (@ 633 nm) on the substrate bias voltage for rf power densities of 2.5 and 5 W/cm^2.

The difference between films sputtered with and without bias is mainly due to incorporation of argon in the films. Kennedy [25] found that application of a bias voltage increases the argon content by one order of magnitude, which is in line with our findings. Films sputtered with substrate bias are more porous and thus have a lower refractive index. Furthermore, they show a greater dependence on process conditions than dense films sputtered without bias.

2.3.3 Film structure and solubility

The structure of vacuum-deposited aluminum oxide films is strongly dependent on the substrate temperature during deposition. Films deposited at low temperatures tend to be amorphous, films deposited at high temperatures polycrystalline. The threshold temperature depends on the process applied. Kennedy [25] obtained polycrystalline films by rf sputtering at 450 °C substrate temperature, while at 350 °C the films were amorphous. Goranchev [21] performed a systematic study into the effects of substrate temperature on the film structure for dc reactive sputtering and found that the transition from amorphous to polycrystalline film structure occurs between substrate temperatures of 290 and 300 °C (at a sputtering pressure around 10 Pa). For evaporated films Bunshah and Schramm [26] found much higher transition temperatures. These results suggest that the film structure is determined by the surface kinetics of the constituent particles during film formation.

Transmission electron microscopy (TEM) and electron diffractometry of zero-bias sputtered films reveal a polycrystalline film structure with a crystallite size in the order of 10 nm. Figure 2.6 shows an electron diffractogram and a dark-field image of a film sputtered with 5 W/cm^2 without substrate bias. The diffractogram corresponds to γ-Al_2O_3. The film structure closely resembles the structure found by Goranchev [21] for polycrystalline films. We found no effect of the anneal treatment on the film structure.

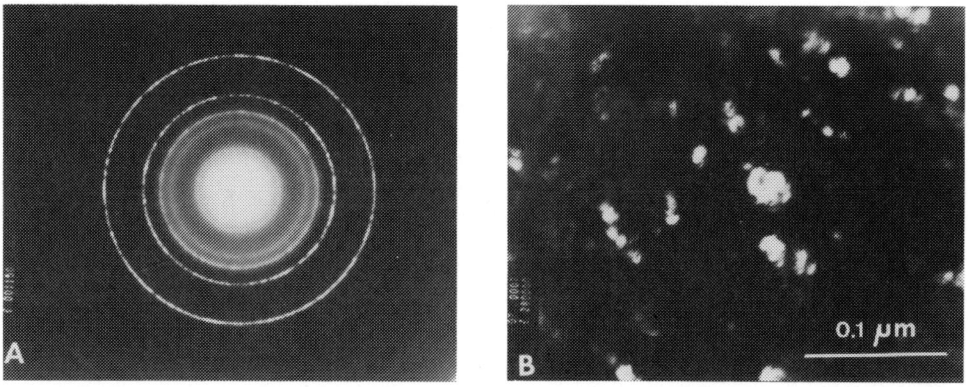

Figure 2.6 *Electron diffraction pattern (**A**) and dark field image (**B**) of an aluminum oxide film sputtered with 5 W/cm*^{2} *without substrate bias. (Courtesy: C. de Haan, Laboratory of Metallurgy, Delft University of Technology.)*

Amorphous aluminum oxide films are soluble in a HF solution whereas compact polycrystalline films are not. We found that films with a refractive index lower than 1.63 are damaged by a buffered HF etch, whereas films with indices higher than 1.67 are not. Film damage is most probably due to a combination of capillary etching of small pores in the aluminum oxide film and under-etching of the silicon dioxide substrate layer through these pores. Our results are in good agreement with those of Hashimoto [27], who found an inverse linear relation between etch rate and refractive index for amorphous films in a 5% HF solution, with zero etch rate occurring at a film index of approximately 1.67.

2.4 Annealing mechanism

Figure 2.7 shows that there is a clear relation between the refractive index, the minimal attenuation A_{min}, and the optimal anneal temperature T_{min}. In addition, there is a relation between the anneal process and film stress. Aluminum oxide films sputtered without substrate bias exhibit an extremely high compressive stress. Measurement of the bending radius of specially prepared substrates after sputtering an aluminum oxide film yield stress values between 2 and 3 GPa. This stress disappears after an anneal treatment.

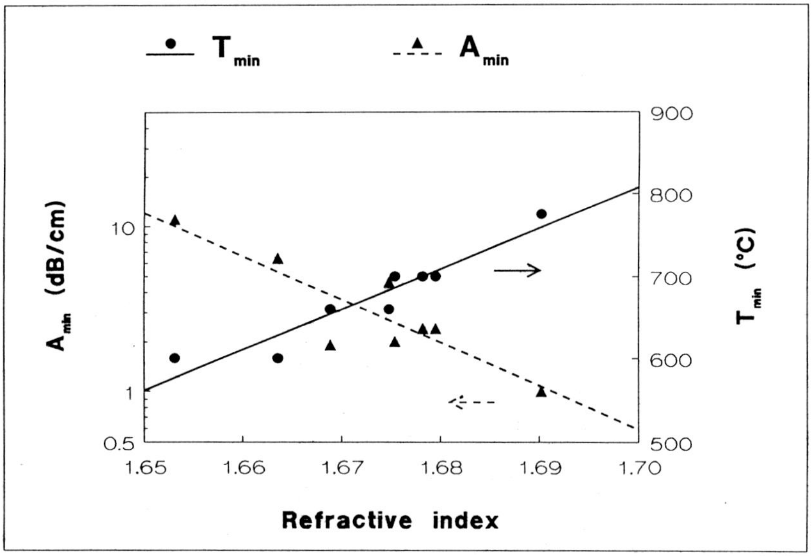

Figure 2.7 The relation between the refractive index, the minimal attenuation (both @ 633 nm), and the optimal anneal temperature, for 0.25-μm films.

Desorption measurements in vacuum on sputtered films which are heated at a rate of 2 K/s show argon desorption peaks at specific temperatures. Figure 2.8 shows the desorption spectra of two films sputtered with 2.5 W/cm², with and without substrate bias. The bias-sputtered film shows a high desorption peak at a temperature of 600 °C, the zero-bias film a much lower peak at 1100 °C. The desorption data agree with results obtained with EDS (Energy Dispersive Spectrometry) which reveal a small amount of argon (<1%) in films sputtered without bias, and a larger amount (a few percent) in bias sputtered films.

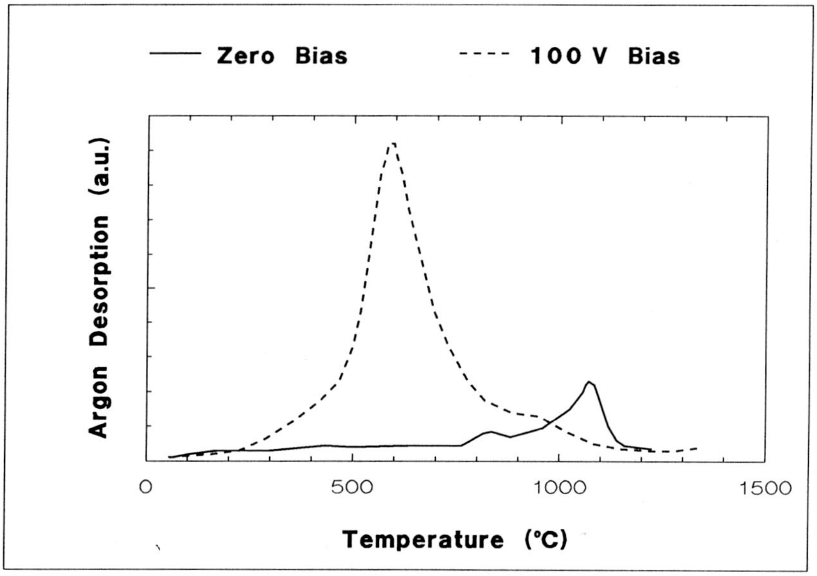

Figure 2.8 Desorption spectra of aluminum oxide films sputtered with 2.5 W/cm² rf power density, with and without bias. The substrates are heated with 2 K/s. (Courtesy: A. van Veen, Interfaculty Reactor Institute (IRI), Delft University of Technology.)

For the film sputtered without bias, desorption begins at a temperature of approximately 850 °C. On isothermal heating the desorption spectrum will shift towards lower temperatures by some 200 °C, which means that at an anneal temperature of 800 °C argon outgassing occurs at a considerable rate, and may thus be responsible for the stress relaxation and the accompanying attenuation reduction. This explanation is supported by the fact that the desorption peak for the bias sputtered film is shifted by some 500 K towards lower temperatures. Extrapolation of the relations plotted in figure 2.7 predict for the bias-sputtered film, which has a refractive index of 1.57, a shift of some 600 K, which is close to the observed value.

The coincidence of argon outgassing and stress relaxation suggests that the argon atoms, which are most probably built in at intercrystalline positions, are responsible for the compressive stress. Nowicki et al. [28] found lower stress values (0.3 GPa) for magnetron sputtered aluminum oxide films with refractive indices from 1.61 to 1.66. This is in agreement with the above explanation: in porous films there will be many vacancies in which argon can be built in without causing film stress. Furthermore, in porous films the built-in argon atoms will be less bound than in compact ones, thus explaining the positive correlation between refractive index and required anneal temperature. The negative correlation between the residual attenuation A_{min} and the refractive index value is in agreement with the results reported in the literature [9,17-19], as discussed in subsection 2.2.2.

The mechanism of the attenuation reduction is less clear. Substitution of the average crystallite size (10 nm) in the approximate expression for scattering loss, as given by Unger [29], yields $\alpha \simeq 2.5\ \sigma_n^4\ cm^{-1}$, in which α is the attenuation constant and σ_n the RMS index variation. To explain the experimentally found attenuation of 20 dB/cm ($\simeq$2.5 cm^{-1}) by stress-induced (piezo-optic) anisotropy due to the random crystallite orientation, an index variation $\sigma_n \simeq 1$ has to be assumed, which is unrealistic. The more realistic assumption $\sigma_n \simeq 0.01$, on the other hand, requires a considerable amount of crystallites having a size in the order of 0.5 µm, which was not observed (see figure 2.5b).

Absorption by the built-in argon atoms might be another explanation, but this is not supported by the visual observation that the scattering intensity is reduced after annealing.

The dramatic increase of the attenuation beyond the optimal anneal temperature is better understood than the observed reduction. If the film stress is completely relaxed the aluminum oxide film will experience tensile stress when it is cooled down to room temperature. This is due to the difference in thermal expansion coefficient between aluminum oxide ($5 \cdot 10^{-6}$/K for bulk sapphire) and silicon ($2.5 \cdot 10^{-6}$/K). For vacuum-deposited films with a granular structure, which are very vulnerable to tensile stress, this difference in thermal expansion coefficient is most probably responsible for the damage observed in films annealed at temperatures beyond the optimal temperature.

The optimal combination of annealing temperature and annealing time is, consequently, that for which the residual compressive stress in the film at the annealing temperature is such, that the film stress is reduced to zero when the substrate is cooled down to room temperature. Relaxation of the residual stress below this value leads to harmful tensile stress at room temperature, so that the optimum in the attenuation curve will be rather critical. This is in agreement with experimental observations.

In conclusion, the stress relaxation and the accompanying attenuation reduction which are observed after a heat treatment are most probably due to outgassing of argon which is incorporated during the sputtering process. The mechanism of the optical attenuation is not yet understood. The film damage observed at annealing temperatures beyond the optimal temperature is due to a difference in thermal expansion coefficients.

Although it is not fully understood, the reproducibility of the annealing process is quite adequate for device research. The experimental work necessary to gain further insight in the annealing mechanism was, therefore, given a low priority.

References

2.1 R.G. Hunsperger, Integrated Optics: Theory and Technology, Chapter 4, pp. 47-69 (Springer: Berlin, Heidelberg, New York, Tokyo, 1984).

2.2 J.E. Goell and R.D. Standley, Sputtered glass waveguide for integrated optical circuits, Bell Syst. Tech. J., Vol. 48, pp. 3445-3448, 1969.

2.3 D.H. Hensler, "Optical propagation in sheet and pattern generated films of Ta_2O_5", Applied Optics, Vol. 10, pp. 1037-1042, 1971.

2.4 J.E. Goell, "Barium silicate films for integrated optical circuits", Applied Optics, Vol. 12, pp. 737-742, 1973.

2.5 R.H. Deitch, E.J West, T.G. Giallorenzi, and J.F. Weller, "Sputtered thin films for integrated optics", Applied Optics, Vol. 13, pp. 712-715, 1974.

2.6 R. Th. Kersten, H.F. Malein, and W. Rauscher, "Optical losses of evaporation-deposited dielectric waveguides", Thin Solid Films, Vol. 28, pp. 369-374, 1975.

2.7 W. Stutius and W. Streifer, "Silicon nitride films on silicon for optical waveguides", Applied Optics, Vol. 16, pp. 3218-3222, 1977.

2.8 V.K. Kiselev and V.P. Red'ko, "Fabrication of Al_2O_3 optical waveguide by reactive rf sputtering", Sov. Phys. Tech. Phys., Vol. 24, pp. 518-519, 1979.

2.9 H. Taguchi, Y. Takahashi, and C. Matsumoto, "Optical properties of thin film of zirconium dioxide", Jpn. J. Appl. Phys., Vol. 19, pp. 1417-1418, 1980.

2.10 L.N. Binh, R.P. Netterfield, and P.J. Martin, "Optical waveguides prepared by ion-beam-assisted deposition", Proc. 10th Eur. Conf. on Optical Communications, Stuttgart, 1984 (Berlin: VDE, 1984), pp. 34-35.

2.11 G. Este and W.D. Westwood, "Reactive deposition of low-loss Al_2O_3 optical waveguides by modified dc planar magnetron sputtering", J. Vac. Sci. Technol., Vol. A2, pp. 1238-1247, 1984.

2.12 M. Ishida, I. Katakabe, T. Nakamura, and N. Ohtake, "Epitaxial Al_2O_3 films on Si by low-pressure chemical vapor deposition", Appl. Phys. Lett., Vol. 52, pp. 1326-1328, 1988.

2.13 K. Sawada, M. Ishida, T Nakamura, and N. Ohtake, "Metalorganic molecular beam epitaxy of γ-Al_2O_3 films on Si at low growth temperatures", Appl. Phys. Lett., Vol. 52, pp. 1672-1674, 1988.

2.14 M.K. Smit, C.J. van der Laan, and G.A. Acket, "Al_2O_3 films for integrated optics", Thin Solid Films, Electronics and Optics, Vol. 138, pp. 171-181, 1986.

2.15 M.K. Smit, "Sputtered planar optical waveguides", Sensors & Actuators, Procs S&A Symp. of the Twente University of Technology, Enschede, The Netherlands, Oct. 30-31, 1986 (Deventer: Kluwer Technical Books, 1986), pp. 45-55.

2.16 B. Chapman, Glow discharge processes, sputtering and plasma etching (New York: Wiley, 1980).

2.17 C.A.T. Salama, "RF sputtered aluminum oxide films on silicon", J. Electrochem. Soc., Vol. 117, pp. 913-917, 1970.

2.18 W.D. Westwood, R.J. Boynton, and S.J. Ingrey, "Effect of pressure on the properties of reactively sputtered Ta_2O_5", J. Vac. Sci. Technol., Vol. 11, pp. 381-384, 1974.

2.19 W.M. Paulson, F.S. Hickernel, and R.L. Davis, "Effect of deposition parameters on optical loss for rf-sputtered Ta_2O_5 and Si_3N_4 waveguides", J. Vac. Sci. Technol., Vol. 16, pp. 307-310, 1979.

2.20 B. Goranchev, V. Orlinov, V.Tsaneva, and I. Petrov, "Direct current reactive sputtering of aluminium", Thin Solid Films, Vol. 52, pp. 365-371, 1978.

2.21 B. Goranchev, V. Orlinov, F. Przyborowski, and Chr. Weissmantel, "Structure of aluminium oxide films deposited by dc reactive sputtering", Thin Solid Films, Vol. 70, pp. 111-116, 1980.

2.22 C. Deshpandey and L. Holland, "Preparation and properties of Al_2O_3 films by dc and rf magnetron sputtering", Thin Solid Films, Vol. 96, pp. 265-270, 1982.

2.23 J. H. Angenent, Onderzoek aan planaire optische golfgeleiders (Research into planar optical waveguides), Master's Thesis, Optics Laboratory, Dept. of Applied Physics, Delft University of Technology, 1984.

2.24 Th. J. M. Jongeling, Technologische aspecten van planaire optische golfgeleiders (Technological aspects of planar optical waveguides), Master's Thesis, Optics Laboratory, Dept. of Applied Physics, Delft University of Technology), 1985.

2.25 T.N. Kennedy, "Evaluating RF sputtered Al_2O_3 films for microcircuit fabrication", Electronic Packaging and Production, Vol. 14, pp. 136-141, 1974

2.26 R.E. Bunshah and R.J. Schramm, "Alumina deposition by activated reactive evaporation", Thin Solid Films, Vol. 40, pp. 211-216, 1977.

2.27 H. Hashimoto, L.L. Levenson, H. Usui, I. Yamada, and T.Takagi, "Optical and structural characteristics of Al_2O_3 films deposited by the reactive ionized cluster beam method", J. Appl. Phys., Vol. 63, pp. 241-244, 1988.

2.28 R.S. Nowicki, "Properties of rf-sputtered Al_2O_3 films deposited by planar magnetron", J. Vac. Sci. Technol., Vol. 14, pp.127-133, 1977.

2.29 H.G. Unger, Planar optical waveguides and fibres (Oxford: Clarendon Press, 1977), p. 26.

Chapter 3

Waveguides

In this chapter the computation of mode properties, such as effective index, attenuation and effective width, is briefly recapitulated. The choice of the waveguide parameters such as film thickness, etch depth, waveguide width, and the thickness of buffer and cover layers, is discussed. Some attention is paid to propagation-loss mechanisms.

3.1 Introduction

Waveguides are basic elements in planar optical circuits. Both passive components such as directional couplers, power splitters, tapers, optical filters, and (de)multiplexers, and active components such as lasers, modulators and polarization controllers are based on optical waveguide structures. The fabrication of well-defined waveguides with low losses is therefore a crucial step in integrated optics technology.

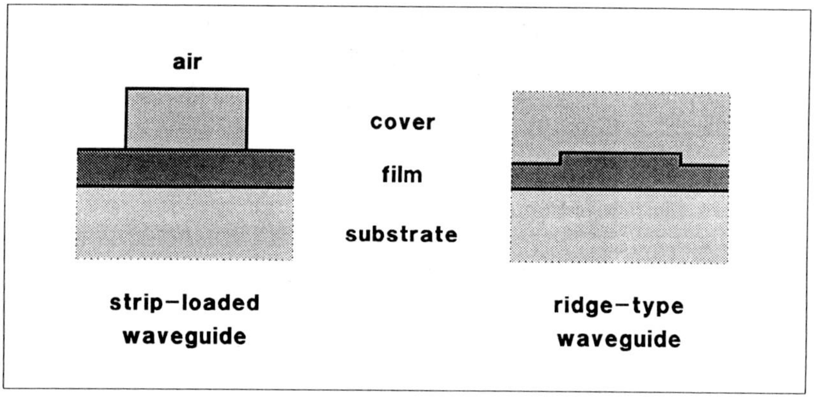

Figure 3.1 The applied waveguide structures.

Lateral guiding can be obtained by creating a refractive index difference between the waveguide and its environment, for example by local implantation or in-diffusion of index-increasing atoms or ions, or by physically shaping the waveguide channel by etching away part of its environment. With the first method it is difficult to produce high index contrasts, with the latter one the whole range from very low to very high contrasts is obtainable. It was therefore preferred for the present research.

Historically, we began with the so-called strip-loaded waveguide (figure 3.1), which was obtained by sputtering an Al_2O_3 and a SiO_2 film onto a thermally oxidized silicon substrate, as described in the previous chapter, and then etching the waveguide pattern into the top SiO_2 layer with a buffered HF etch. This waveguide structure is easily fabricated, but appeared to have some major drawbacks:

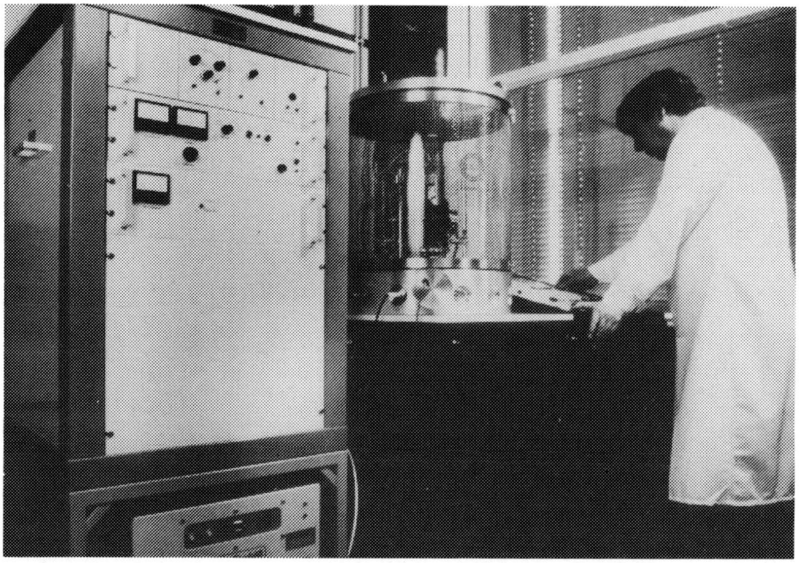

Figure 3.2a The atom-beam milling machine for etching the ridge waveguides.

Figure 3.2b Close-up of the gun (Ion Tech FAB 92 Flex) which produces the accelerated argon-atom beam, and the substrate table.

- The edge roughness of the strip, which is due to the microstructure of the sputtered film, introduces scattering losses. For 3 µm wide waveguides the lowest additional loss obtained amounted to 12 dB/cm (@ 633 nm), which is unacceptably high.
- With a 0.6 µm thick SiO_2 top layer, necessary for prism coupling, the underetch (single-sided) varied between 0.6 and 1.0 µm; this is prohibitive for producing narrow waveguides. At the longer wavelengths the problem becomes even greater due to the thicker top layer required.
- The strip-loaded guide is sensitive to contamination because part of the light flux travels through the uncovered region next to the strip.
- If a prism is contacted to the strip guide with a high-index matching liquid, as described in chapter 8, the liquid produces a lateral anti-guide, thus impeding coupling of light into narrow waveguides.

All of these drawbacks are overcome by applying a ridge-type waveguide structure as depicted in figure 3.1. The ridge is obtained by milling the Al_2O_3 film with accelerated argon atoms ($\simeq$1keV) through a photo-resist mask with the etching machine shown in figure 3.2. The mask is removed in acetone, carbon residues produced during the bombardment are removed by a 10-minutes bake at 700 °C under air. The whole structure is covered with a rf-magnetron sputtered SiO_2 film.

Scattering losses are lower by more than one order of magnitude compared with the strip-loaded guide. The lateral degradation of the mask during the etch is less than 0.1 µm (single-sided). Prism-coupling losses are close to the theoretically expected values. Finally, the lateral contrast can be chosen accurately along a wide range of values by controlling the etch depth through the etching time.

A disadvantage of the ridge-type waveguide is the more laborious fabrication procedure. The advantages, however, are so pronounced that we ceased to employ the strip-loaded guide. In the next sections the choice of the waveguide parameters will be discussed.

3.2 Computation of mode properties

Figure 3.3 shows the geometry of the applied waveguides, which consist of a five-layer structure:

- a silicon substrate,
- a thermal silicon-dioxide buffer layer which has to be thick enough to isolate the guided light optically from the high-index silicon substrate and thus serves as a transparent substrate,
- the aluminum oxide waveguide layer,
- a sputtered silicon-dioxide cover layer,
- air superstrate.

In many computations the structure is considered to be a three-layer structure, with the buffer layer a substrate with refractive index n_s, and the cover layer a superstrate with refractive index n_c. The film has a thickness d and a refractive index n_f. The ridge guide is created by etching a step Δd in this film. We shall first consider the computation of mode properties in a two-dimensional waveguide structure.

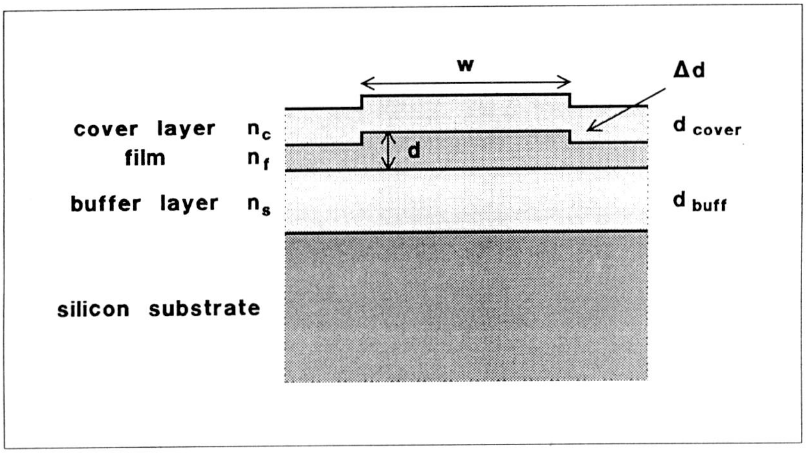

Figure 3.3 Ridge-guide configuration.

3.2.1 Two-dimensional waveguide structures

A. Effective index in lossless three-layer structures

The most elementary parameter in describing the properties of a waveguide mode is its propagation constant β, or the effective index *N* which is related to it according to

$$N = \beta/k_o, \tag{1}$$

k_o being the propagation constant (or wave number) in vacuum. The effective index of the mode depends on the parameters of the waveguide structure through the well-known dispersion relation. The most important formulae for determining N will be briefly recapitulated. In its normalized form the dispersion equation for a lossless three-layer structure is written as [1,2]:

$$V\sqrt{1-b} - arctan\left\{\left(\frac{n_f^2}{n_s^2}\right)^q \sqrt{\frac{b}{1-b}}\right\} - arctan\left\{\left(\frac{n_f^2}{n_c^2}\right)^q \sqrt{\frac{b+a}{1-b}}\right\} = m\pi, \tag{2}$$

with $q=0$ for TE-modes and $q=2$ for TM-modes. V is the normalized film parameter:

$$V = k_o d \sqrt{n_f^2 - n_s^2}. \tag{3}$$

The integer m is the mode number, for the fundamental mode it equals zero. The parameter a is the so-called asymmetry parameter:

$$a = \frac{n_s^2 - n_c^2}{n_f^2 - n_s^2}. \tag{4}$$

For symmetrical waveguides it equals zero. The parameter b is the normalized propagation constant:

$$b = \frac{\beta^2 - n_s^2 k_o^2}{n_f^2 k_o^2 - n_s^2 k_o^2} = \frac{N^2 - n_s^2}{n_f^2 - n_s^2}, \qquad 0 < b < 1, \tag{5}$$

with the following inverse relation:

$$N = \sqrt{n_s^2 + b(n_f^2 - n_s^2)} \simeq n_s + b\,(n_f - n_s). \tag{6}$$

The latter approximation holds for small values of the relative refractive index contrast Δ:

$$\Delta = \frac{n_f - n_s}{n_s}. \tag{7}$$

For *TE*-modes the normalized propagation constant b depends on the film parameters only through V and a, so that a normalized solution of the dispersion relation is possible. Figure 3.4 shows this solution for the first two modes, for a number of different asymmetries.

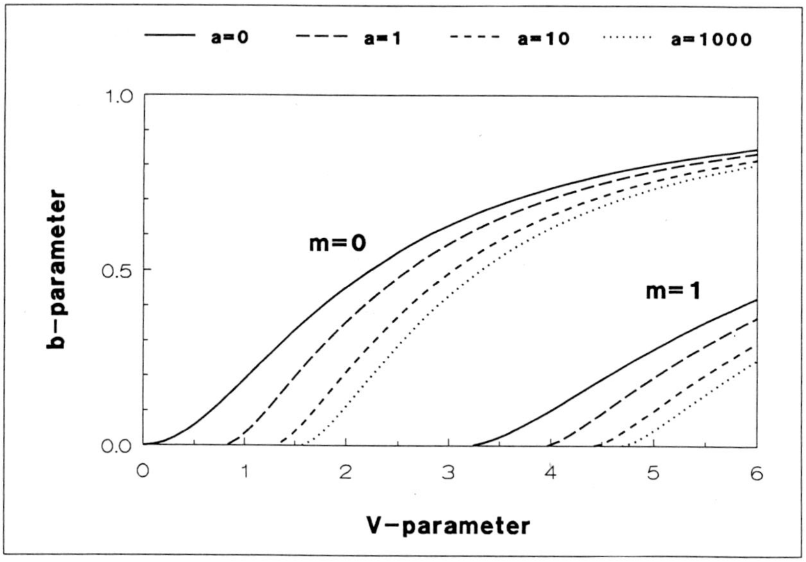

Figure 3.4 *The normalized b-V-diagram*

For small values of Δ the normalized solutions also apply to *TM*-modes if the asymmetry parameter is modified as follows:

$$a_{TM} = \left(\frac{n_f}{n_c}\right)^4 a_{TE}, \tag{8}$$

in which a_{TE} is the asymmetry parameter as defined in equation 3.4. For the aluminum oxide / silicon dioxide waveguide system $\Delta \simeq 0.16$, so that the approximation is very inaccurate.

B. Effective mode width

The effective width of a mode appears in many formulae. A simple expression for the effective width of the fundamental mode will be derived which can also be employed for approximation of the mode profile as a Gaussian beam. This approximation allows for a quick estimation of the far-field pattern.

The effective mode width is defined as the width of a uniform intensity distribution having the same maximum intensity and the same power content as the mode profile $U(x)$:

$$w_e = \frac{\int U^2(x)\,dx}{U^2_{max}}. \tag{9}$$

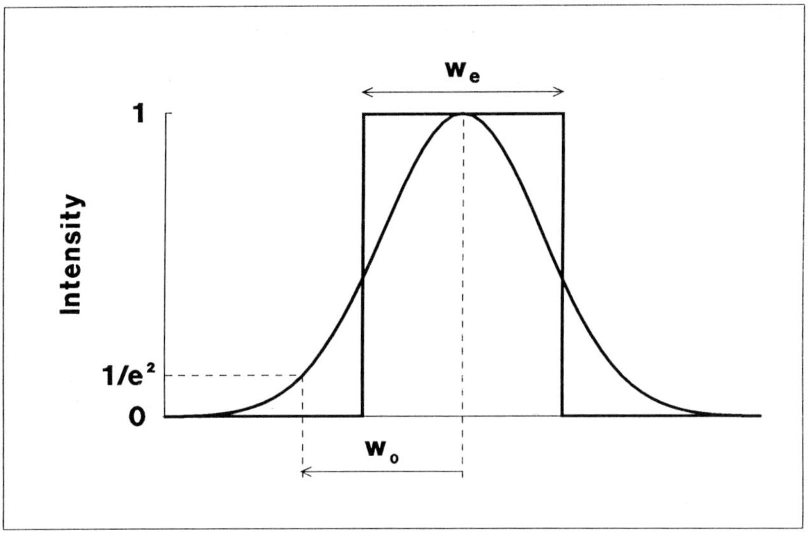

Figure 3.5 *The effective mode width as the width of a uniform distribution with the same maximum intensity and power content as the mode profile.*

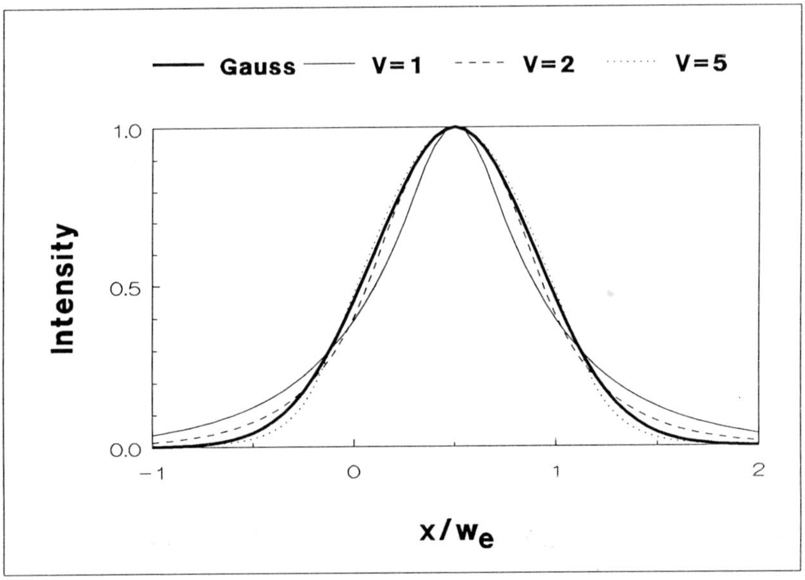

Figure 3.6 *Mode profiles of TE-modes with equal effective widths and V-parameters 1, 2, and 5 and their Gaussian approximation according to formula 3.10.*

The concept of the effective width is illustrated in figure 3.5. For a Gaussian beam profile $exp(-x^2/w_o^2)$ we find an effective width $w_o\sqrt{(\pi/2)}$, so that a mode can be approximated by a Gaussian beam with beam waist:

$$w_o = w_e\sqrt{(2/\pi)}. \tag{10}$$

Figure 3.6 shows the quality of this approximation for *TE*-polarized modes with V-parameters of 1, 2, and 5. For modes in three-layer waveguides relation 3.9 can be computed analytically:

$$w_e = \tfrac{1}{2}w\,(1 + \frac{1}{v} + \frac{1}{\underline{w}}), \tag{11}$$

in which w is the waveguide width or film thickness, and v and $\underline{w}$ are the normalized transverse attenuation constants in the substrate and the cover layer,respectively. These are, together with the normalized propagation constant u in the film, defined as[1] [2]

$$u = k_o d\sqrt{(n_f^2-N^2)}, \tag{12a}$$
$$v = k_o d\sqrt{(N^2-n_s^2)}, \tag{12b}$$
$$\underline{w} = k_o d\sqrt{(N^2-n_c^2)}. \tag{12c}$$

Formula 3.11 is valid for *TE*-polarized modes, for *TM*-polarized modes it can be used as an approximation if the index contrast is small. The effective width can be approximated by the following empirical formula:

---

1) The normalized attenuation constant $\underline{w}$ is underlined in order to distinguish it from the waveguide width w.

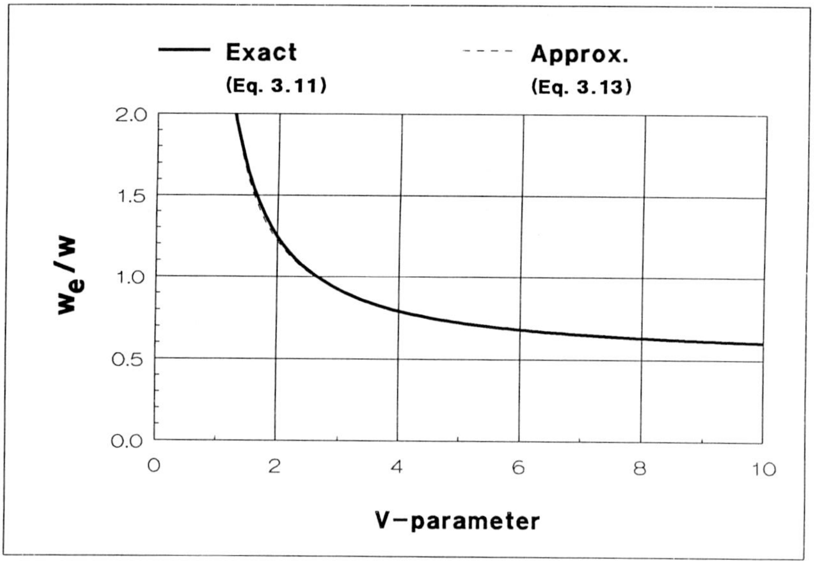

Figure 3.7 The ratio w_e/w (for TE-modes) as a function of the V-parameter.

$$w_e = w\,(0.5 + \frac{1}{V - 0.6}). \qquad (13)$$

Figure 3.7 shows both formula 3.11 and its approximation (formula 3.13).

C. Effective index in absorbing or multi-layer structures

The dispersion relation, as described in subsection A, only applies to three-layer lossless waveguide structures. For planar waveguide structures with more than three layers, or with complex refractive indices the propagation constant may become complex. A complex propagation constant γ is defined according to:

$$\gamma = \alpha + j\beta, \tag{14}$$

in which β has the same interpretation as in lossless waveguides, and the attenuation coefficient α describes the exponential attenuation of the modal field in the propagation direction. If the corresponding complex effective index is written as:

$$N = N' - jN'', \tag{15}$$

then the attenuation coefficient α is related to N'' as

$$\alpha = N''k_o. \tag{16}$$

For this type of waveguide we applied a transfer-matrix method [3] to determine the complex values N corresponding to the waveguide modes. Our method supports the computation of effective indices both for guided and leaky modes, the latter under the restriction that the attenuation coefficients are small, i.e. $N'' \ll N'$.

3.2.2 Three-dimensional waveguide structures

The effective index method

Calculations on three-dimensional waveguide structures were performed using the Effective Index Method (EIM), as introduced by Knox and Toulios [6] for dielectric microwave guides. In this method, which is illustrated in figure 3.8, the waveguide is separated into a number of regions in which the structure is invariant in the lateral direction. In each of these regions the effective index *N* is computed as if the region were infinitely extended in the lateral direction (see also [7-9]). Next, the propagation constant of the waveguide is computed by considering it as a two-dimensional waveguide in which each region is represented as a film having the lateral

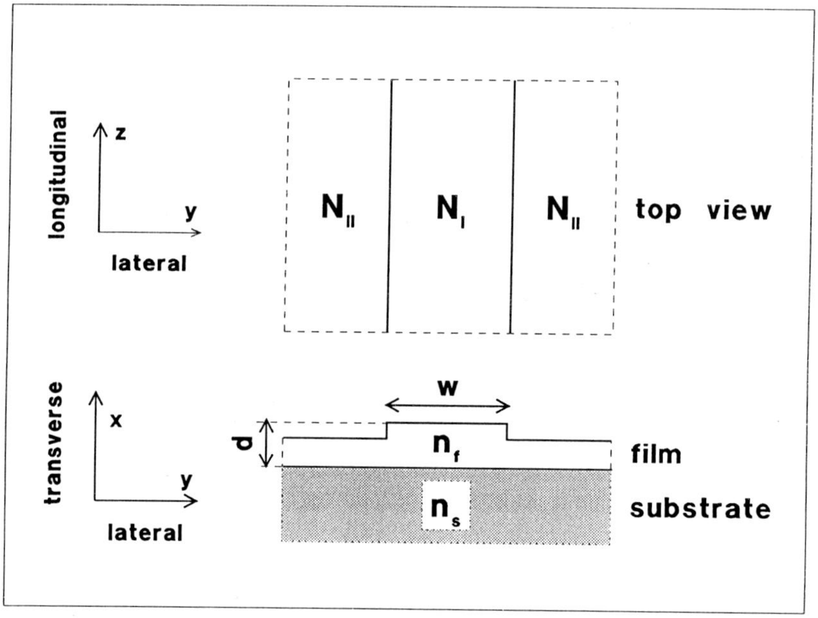

Figure 3.8 The Effective Index Method illustrated for a ridge-guide geometry.

width of the region as film-thickness and its effective index as actual refractive index.

Three-dimensional waveguide modes will be annotated as TE_{mn} or TM_{mn} according to the nomenclature proposed by Baken [10]. In this notation the polarization of a mode refers to that of the corresponding slab mode which is obtained by infinitely increasing the waveguide width. The indices m and n refer to the transverse and the lateral mode number, respectively.

Baken [11] analyzed the waveguide structure, illustrated in figure 3.3, with the Domain-Integral Method for different etch depths Δd. In the case

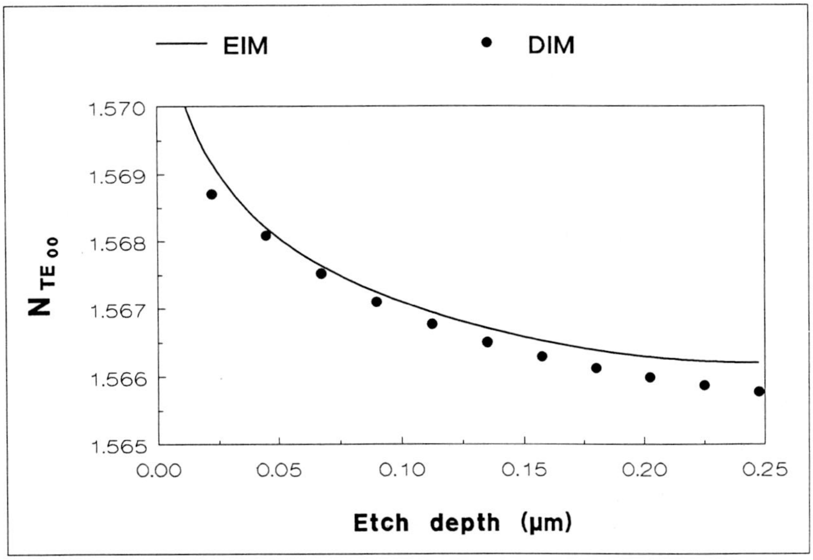

Figure 3.9 *Effective index results for a 2 μm wide ridge guide (with n_f=1.69, $n_s = n_c$=1.457, d=0.25 μm, @633 nm wavelength) for varying etch depth, as computed with the Effective Index Method and the Domain-Integral Method (Baken [11]).*

of the fundamental mode the differences between this method and the Effective Index Method are within $5 \cdot 10^{-3}$ of the lateral contrast for 2 µm waveguides and ridge heights Δd up to half the film thickness, as can be seen from figure 3.9. For both very high and very low contrasts the discrepancy increases. As the Effective Index Method is based on separation of variables, it will become less accurate if the transverse mode profiles in and next to the ridge differ considerably. This will happen where etch depths are large. For small etch depths the method is expected to be reliable for the same reason. The discrepancy for small etch depths is produced by the DIM because of the small aspect ratio of the domains which causes an inaccurate description of the electromagnetic field in the ridge region.

For many practical purposes the accuracy of the Effective Index Method is sufficient. In the following sections the choice of the waveguide parameters will be discussed.

3.3 Al_2O_3 film thickness

From figure 3.4 it may be seen that the steepest part of the *b*-*V*-curve falls between *V*-parameter values of 1 and 2. In this range a lateral effective index contrast of about 25% of the transverse contrast is obtained by etching away 50% of the film. This contrast is sufficient for many applications. In order to minimize the variation in process parameters we standardized the Al_2O_3 film thickness to the *V*=2 value for the four wavelengths tabulated in table 3.1.

λ_0	$d_{V\approx 2}$	$n_{Al_2O_3}$	n_{SiO_2}	N_{TE}	N_{TM}
633 nm	0.25 µm	1.690[1]	1.457[1]	1.573	1.553
820 nm	0.30 µm	1.683[1]	1.453[2]	1.559	1.539
1300 nm	0.50 µm	1.677[1]	1.447[2]	1.558	1.538
1550 nm	0.60 µm	1.673[1]	1.444[2]	1.555	1.535

Table 3.1 Film thickness corresponding to V≃2, and the refractive index values, on which it is based.

1) Measured (as described in chapters 7 and 8).
2) Inferred from references 4 and 5.

3.4 Ridge height and waveguide width

If we choose the film thicknesses corresponding to the wavelengths listed in table 3.1 such that the V-parameter is the same for all wavelengths *(V≃2)*, then the lateral effective index contrast as a function of the relative ridge height δd:

$$\delta d = \Delta d/d, \tag{17}$$

is independent of wavelength and polarization within a few percents. Figure 3.10 shows the dependence for *TE*- and *TM*-polarization. These curves can, therefore, be used through the whole wavelength range from 633 to 1550 nm.

Figure 3.11 shows the width below which the waveguide is monomode. For the long wavelengths it is relatively easy to keep waveguides monomode.

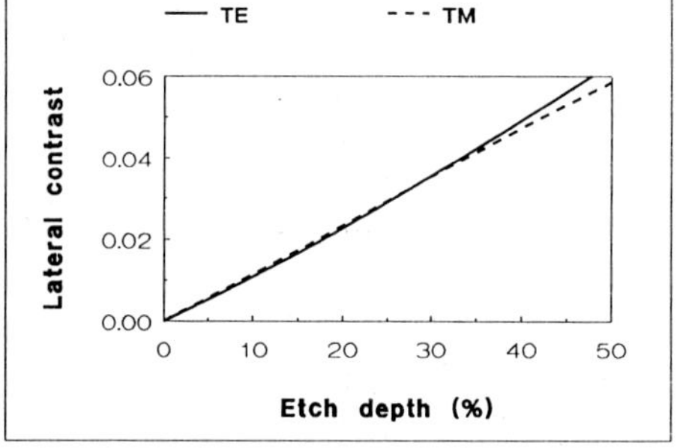

Figure 3.10 *Lateral effective index contrast as a function of the relative edge depth for the waveguide structures listed in table 3.1.*

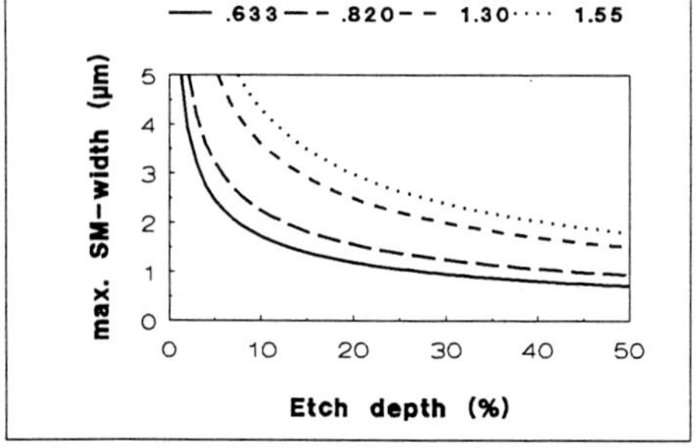

Figure 3.11 *Maximum waveguide width for monomode operation, as a function of the relative etch depth.*

2-μm waveguides remain monomode up to relative ridge heights of 30% (1300 nm) and 45% (1550 nm), corresponding to a lateral contrast of 0.04 and 0.06 respectively. For the shorter wavelengths, however, they become multi-moded already at relative ridge heights of 7% (633 nm) and 12% (820 nm), corresponding to lateral contrasts in the order of 0.01.

Waveguides smaller than 2 μm are difficult to fabricate reproducibly with optical lithography. If a high contrast is required it will, therefore, be necessary to apply multimode waveguides. For the research reported in the present thesis, 2 μm was adopted as a standard waveguide width.

3.5 Cover-layer thickness

An important advantage of the embedded ridge guide is that the cover layer can be used as a tunneling layer for evanescent-field coupling of light into or out of the waveguide (prism coupling), as described in chapter 8. We applied a prism of SF13 glass in combination with CH_2I_2 as an index matching liquid (see table 3.2).

For optimal coupling of a Gaussian laser beam into a waveguide the coupling length of the prism-coupling configuration has to amount to 1.5 times the width of the laser beam measured along the bottom of the prism (see chapter 8) which is in the order of 1 mm in the experimental measurement configurations used in the author's laboratory. The coupling length l_c for a

λ_0	n_{SF13}	$n_{CH_2I_2}$
633 nm	1.763	1.74
820 nm	1.722	
1300 nm	1.709	
1550 nm	1.705	

Table 3.2 Refractive index of SF13 glass and CH_2I_2 (data from references 5 and 12; the refractive index of the applied SF13 prisms was experimentally verified at 633 nm wavelength [13]).

mode follows from the attenuation coefficient α in a prism-loaded (four-layer) waveguide structure:

$$l_c = 1/\alpha. \tag{18}$$

The attenuation coefficient can be computed as discussed in subsection 3.2.1C. Figure 3.12 shows the relation between the coupling length and the relative cover layer thickness d_{cover}/d_{film} for the waveguide structures and wavelengths listed in table 3.1. The relation is slightly wavelength dependent. For the short wavelengths a ratio of 2.4 yields a coupling length between 1 and 2 mm both for the fundamental *TE* and *TM*-mode. For the long wavelengths a ratio of 2.25 yields the same result. We thus find optimal cover-layer thicknesses of 0.6, 0.8, 1.15, and 1.35 µm, for 633, 820, 1300, and 1550 nm wavelengths respectively.

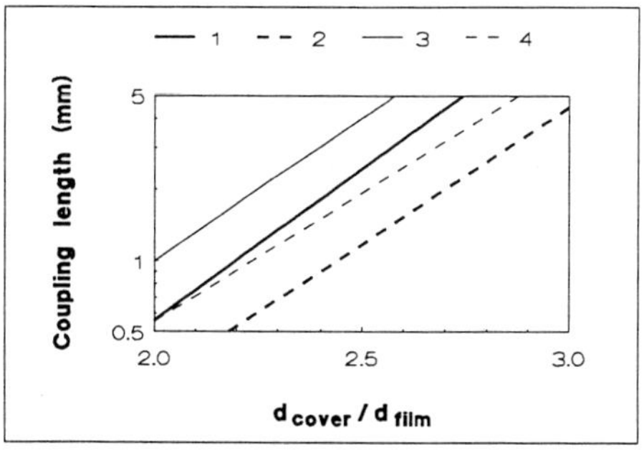

Figure 3.12 *Dependence of the coupling length on the relative cover layer thickness:*

Curves 1,2 : TE, TM at 633 nm
Curves 3,4 : TE, TM at 1550 nm

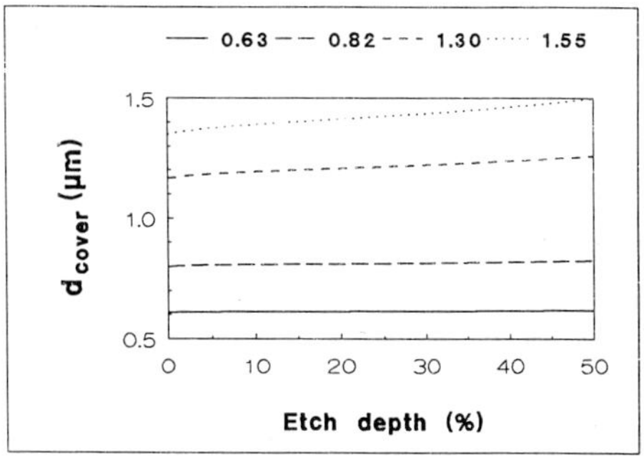

Figure 3.13 *Dependence of the cover-layer thickness, required for a 2-mm coupling length (TE) on the relative etch depth, for the standard waveguide structures listed in table 3.3.*

The coupling length for a ridge-guide mode will decrease with increasing relative etch depth δd due to the reduced transverse confinement of the light. The effect can be computed by applying the Effective Index Method to the complex effective indices of the prism-loaded waveguide structures in and next to the ridge region and inferring l_c from equations 3.18 and 3.16. Figure 3 .13 shows the increase of the cover-layer thickness, which is required to keep the coupling length at a constant value of 2 mm for the TE_0-mode, as a function of the relative etch depth. For the shorter wavelengths the variation is negligible for most practical purposes. For the longer wavelengths the effect is greater due to the smaller lateral confinement (smaller V-parameter) of the mode. If the cover layer thickness is not corrected for this effect, the coupling length will be reduced by a factor of 2 for the worst case shown in figure 3.13 (1550 nm, 50 %). Numerical evaluation shows that this will lead to a 10% decrease in the coupling efficiency.

3.6 Substrate buffer layer thickness

Due to the high refractive index of the silicon substrate, some leakage of guided light to the substrate will occur. The leakage is primarily determined by the thickness of the thermal oxide buffer layer. It is only weakly dependent on the refractive index (both the real and the imaginary part) of the substrate. To keep the radiation loss low the thermal oxide buffer layer must be sufficiently thick.

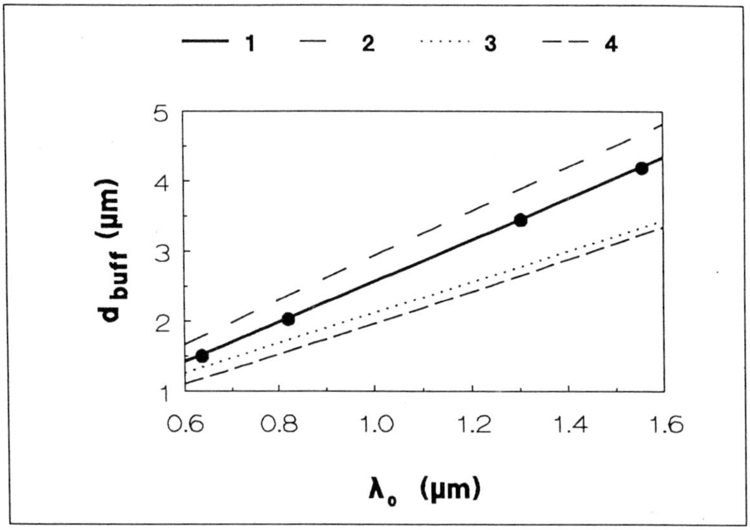

Figure 3.14 Buffer-layer thickness required for low substrate radiation. The bold line represents the most relevant case (the wavelengths listed in table 3.1 are indicated with dots, A_{TM} : radiation loss of TM_{00}-mode)

Curve 1: A_{TM}=0.1 dB/cm, δd = 50%, w=2 µm
*Curve 2: A_{TM}=0.1 dB/cm, δd = 50%, **w=1 µm***
*Curve 3: A_{TM}=0.1 dB/cm, **δd = 25%**, w=2 µm*
*Curve 4: A_{TM}=**1.0 dB/cm**, δd = 50%, w=2 µm*

Figure 3.14 shows the buffer-layer thickness required for keeping the radiation losses for the TM_{00}-mode below 0.1 dB/cm (TE_{00}-losses are consistently lower than TM_{00}-losses), as a function of the wavelength. The waveguide structure was adapted to the changing wavelength in such a way as to keep the transverse V-parameter at a value of 2 and the *(TE)* coupling length on 2 mm (see sections 3.3 and 3.4).

Based on this figure we chose 2 µm as a minimum buffer-layer thickness for the shorter wavelengths and 4 µm for the longer ones. Thick oxide layers have two disadvantages: oxidation times (with conventional atmospheric-pressure steam oxidation) become very long and the substrates cannot be cleaved in a controllable way because the compressive stress in the oxide film impedes cleave propagation. For the short wavelengths 2-µm layers are, therefore, preferred.

3.7 Propagation loss

Propagation loss may be due to the following mechanisms:

* Volume scattering due to film-inhomogeneity.
* Surface scattering at rough interfaces, both transverse (film interfaces) and lateral (waveguide edges).
* Radiation to the substrate (discussed in section 3.5)
* Absorption at color centers.

In the following paragraphs the loss mechanisms which may be responsible for the observed propagation loss will be discussed.

3.7.1 Loss mechanism in slab guides

Volume scatter may occur in both the waveguiding film and the cladding layers. The attenuation coefficient α of a mode due to volume scattering can be written as:

$$\alpha = \eta_f \alpha_f + \eta_s \alpha_s + \eta_c \alpha_c, \tag{19}$$

in which α_f, α_s, and α_c are the attenuation coefficients of bulk material. Over the course of many experiments we found no evidence for significant polarization dependence of these entities. The weight coefficients η_f, η_s, and η_c depend on both the waveguide structure and the type (polarization and order) of the propagating mode. Expressions for the weight coefficients are derived in the appendix at the end of this chapter.

Figure 3.15 shows the dependence of $\eta_f \alpha_f$ and $\eta_c \alpha_c$ (curves 1 and 2) on the film thickness *d*, for the aluminum oxide waveguide system at a wavelength of 633 nm. The bulk attenuation α_s of the thermal oxide is assumed to be negligible. Waveguide attenuation was determined experimentally for films with 0.125, 0.25 and 0.58 µm thickness. All curves in the figure are calibrated with respect to the experimentally found attenuation of 1 dB/cm for a 0.25 µm thick film.

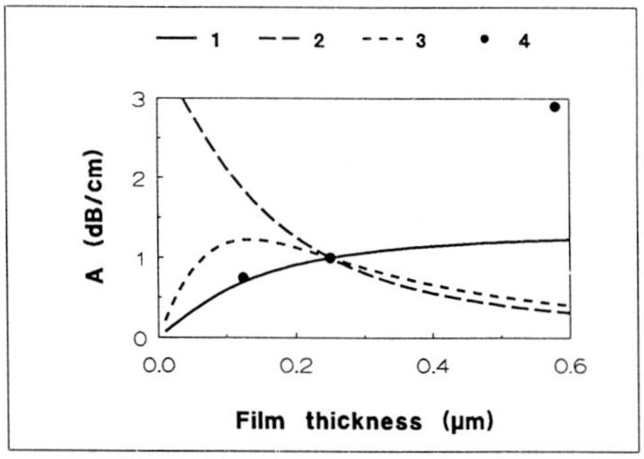

Figure 3.15 Loss contributions of different scatter mechanisms (calibrated on 1 dB/cm at d=0.25 μm):

1 Volume (film)
2 Volume (cover)
3 Surface (film/cover interface)
4 Measured

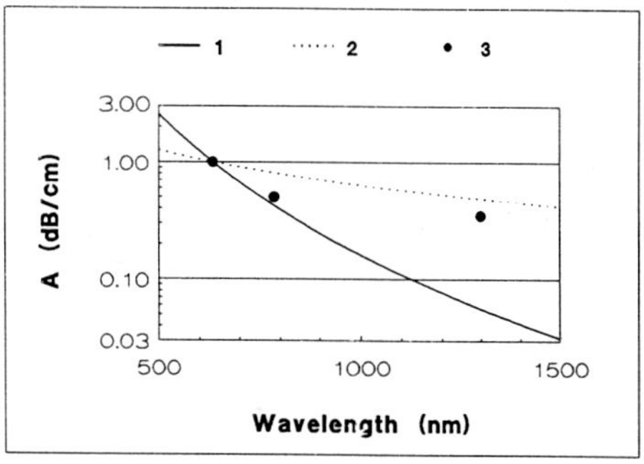

Figure 3.16 Wavelength dependence of the attenuation for the standard waveguide structure

1 Volume scatter
2 Surface scatter
3 Measured

From the figure it may be seen that the increase in attenuation from 0.125 to 0.25 μm film thickness is fully explained by the volume scatter mechanism in the film. The anomalously high value measured for the 0.58 μm film is probably due to the fact that thick films require longer annealing times (which was discovered later).

Surface scattering will occur mainly at the film-cover interface, the thermally oxidated substrate layer being extremely flat. Unger [2] derived the following formula for the scattering loss at this interface:

$$\alpha = \frac{2}{3}\pi\sigma_h^2 B_y B_z \frac{n_c k_o^2 V^2}{N d_e d^2}(1-b)(n_f^2-n_c^2)\, h\left(\frac{n_f^2-n_c^2}{n_c^2}\right). \qquad (20)$$

In this formula σ_h^2 is the mean square height variation of the rough surface and B_y and B_z are the autocorrelation distances in the y and the z-direction, respectively. The function h approaches unity for small values of the index contrast; for the aluminum oxide waveguide system it has the value 1.1. The other parameters are as described in section 3.2.

The formula is based on a description of the surface roughness as a series of local perturbations of the ideal waveguide structure, with volume ΔV and dielectric contrast $\Delta\varepsilon = \pm\varepsilon_o(n_f^2-n_c^2)$, the sign depending on the location of the perturbation relative to the interface. Each volume is then considered as a dipole with electric dipole moment $j\omega\Delta\varepsilon E\Delta V$, in which E is the unperturbed modal field strength at the interface (Born approximation). Integration of the average power scattered in all directions (4π steradians) under the assumption of an exponentially distributed surface roughness with directional autocorrelation distances B_y and B_z yields the power decay of the guided mode and hence the attenuation coefficient. By integrating the scatter over 4π steradians the power scattered forwardly within the acceptance angle of the mode is also considered as loss. If B_y or B_z increase relative to the wavelength (in the waveguide) the forward scatter will tend to increase and equation 3.20 will become increasingly pessimistic. For sputtered films B_y and B_z are expected to be small.

Curve 3 of figure 3.15 gives the thickness-dependence according to equation 3.20; it is calibrated on 1 dB/cm at $d=0.25$ μm. It is clear that this phenomenon cannot explain the observed thickness dependence of the attenuation since a reverse trend was observed.

Absorption loss will show the same thickness dependence as volume scattering loss, and thus cannot be distinguished from the above observations. In the near infrared absorption losses will tend to increase with increasing wavelength, however, which was not observed. Volume scatter will decrease with λ^{-4} (Rayleigh's law). For the standard waveguide geometry ($V \simeq 2$ for all wavelengths, i.e. d and d_e scale with λ) the surface scatter can be seen to reduce with λ^{-1} from equation 3.20. Figure 3.16 shows the predicted wavelength dependence of the volume and surface scattering loss together with the experimental data for the standard waveguide geometry. The experimental data support the conclusion that, at short wavelengths, volume scattering is the dominant loss mechanism. At the longer wavelengths surface-scattering seems to gain importance.

3.7.2 Scattering at rough waveguide edges

If most of the radiation scattered by rough waveguide edges is captured into the slab next to the ridge the scattering problem may be treated as a two-dimensional one. For this case Unger [2] derives a formula which differs only slightly from the three-dimensional one:

$$\alpha = \sigma_h^2 B_z \frac{k_o V^2}{N w_e w^2} (1-b)(N_1^2 - N_2^2), \tag{21}$$

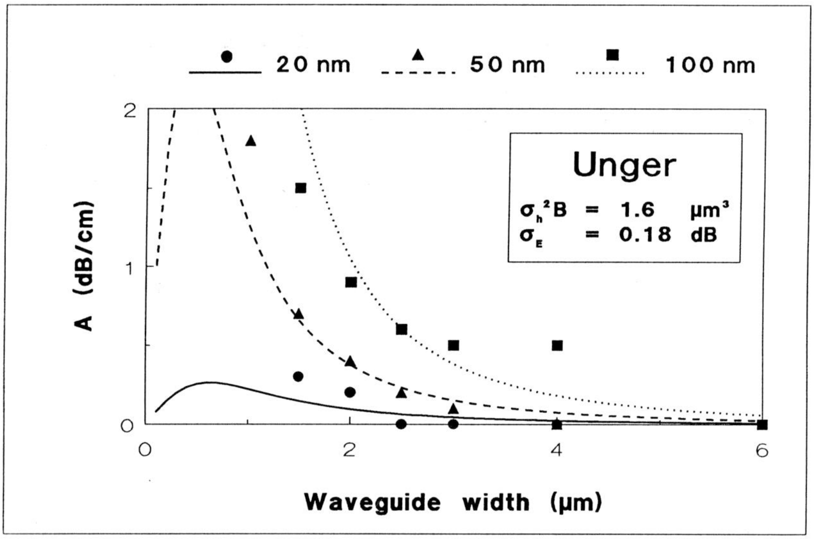

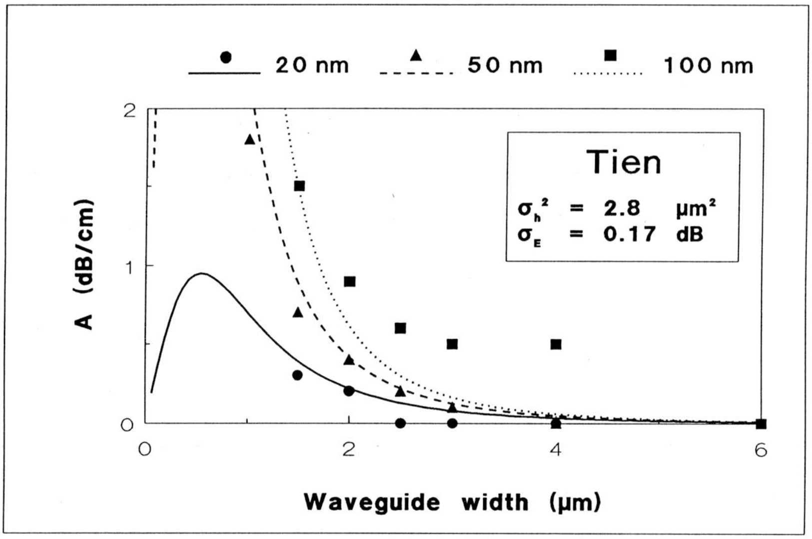

Figure 3.17 *Scattering loss at rough waveguide edges according to the formulas of Unger (equation 3.21, figure a) and Tien (equation 3.22, figure b) for 2-µm wide ridge guides with 20, 50 and 100 nm etch depth (@633 nm). Lines indicate the theoretical predictions, dots the experimental data.*

in which N is the effective mode index, and N_1 and N_2 are the transverse effective indices in and next to the ridge region, respectively (see figure 3.8). It has been assumed that both edges make an equal and uncorrelated contribution to the total loss. The scale of the edge roughness generally being in the order of the wavelength, some prudence is required in the employment of the above formula. Deri et al. [15,16] claim a good fit of the experimental data to a simple formula, derived by Tien [17]:

$$\alpha = 4\sigma_h^2 \frac{n_f^2 k_o^2}{d_e} \frac{\sin^3 \theta}{\cos \theta}, \tag{22}$$

in which θ is the propagation angle of the ray corresponding to the mode under consideration (measured relative to propagation axis) and d_e is them effective film thickness. The formula is based on the scattering theory after Lord Rayleigh, which states that the power carried by a beam is attenuated on reflection at a rough surface by a factor of $exp(-4n_f^2 k_o^2 \sigma^2/\sin^2\theta)$. Equation 3.22 follows by calculating the number of reflections per unit length and multiplying it with the loss per reflection. Marcuse's calculations [18] show that the above reflection formula only applies to the case of a large autocorrelation distance. Because the formula of Unger does not apply to this range, these formulae seem to be more or less complementary.

We tested both formulae by fitting them to a series of attenuation measurements carried out at 633 nm, on waveguides with three different index contrasts *($\Delta n \simeq 0.01, 0.02$ and 0.05)*, and seven different widths *(1, 1½, 2, 2½, 3, 4,* and *6 µm)*. Figure 3.14 shows the results. In the Unger formula (3.21) the value $\sigma_h^2 B=1.6\ \mu m^3$ provides a best fit, with a standard error of estimate $\sigma_E=0.18\ dB$. Tien's formula (3.22) yields a minimal standard error $\sigma_E=0.17\ dB$ for $\sigma_h^2=2.8\ \mu m^2$.

From these σ_E-values no decision can be made regarding which model provides the best description of the scattering mechanism, although the Unger formula seems to produce a slightly better fit for the data on wider waveguides (2-6 µm) which are considered more reliable than the data for very narrow waveguides. Further, it is noted that an edge roughness in the order of 1.5 µm, as inferred from the Tien formula, is unrealistic. Loss data measured at long wavelengths may give decisive information: Unger's formula predicts a λ^{-3} dependence for the standard waveguide geometry with constant index contrast; the Tien formula predicts a weak wavelength dependence. Such data were not available at the moment of writing.

3.8 Conclusions

Choosing the aluminum oxide film thickness *d* (figure 3.3) such that the transverse *V*-parameter equals 2 yields maximum lateral index contrast with minimal etching time. Lateral contrasts up to 0.06 can be obtained with relative etch depths up to 50%, through the wavelength range from 633 to 1550 nm. A silicon dioxide cover layer of 2.4 times the film thickness for the short wavelengths and 2.25 times for the long ones, provides prism coupling lengths between 1 and 2 mm for the two polarizations, both for slab and 2 µm wide ridge guides. A silicon dioxide buffer layer of 2 µm for 633 and 820 nm, and 4 μm for 1300 and 1550 nm provides a substrate isolation better than 0.1 dB/cm for relative etch depths up to 50%. Table 3.3 summarizes the relevant waveguide parameters.

Waveguide attenuation at the short wavelengths is dominated by volume scatter in the aluminum oxide film (1 dB/cm @633 nm, and 0.5 dB/cm @780 nm). At the long wavelengths film surface scatter seems to become dominant. Additional loss due to edge roughness amounts to maximally 2 dB/cm @633 nm wavelength, for a relative etch depth of maximally 50% *($\Delta n \simeq 0.06$)* and 2 μm waveguide width. At the long wavelengths this loss is expected to be lower.

λ_0	$n_{Al_2O_3}$	n_{SiO_2}	d_{film}	d_{cover}	d_{buffer}
633 nm	1.690[1]	1.457[1]	0.25 μm	0.60 μm	2.0 μm
820 nm	1.683[1]	1.453[2]	0.30 μm	0.80 μm	2.0 μm
1300 nm	1.677[1]	1.447[2]	0.50 μm	1.15 μm	4.0 μm
1550 nm	1.673[1]	1.444[2]	0.60 μm	1.35 μm	4.3 μm

Table 3.3 Material refractive indices and waveguide parameters for realizing waveguides with lateral index contrasts up to 0.06 with relative etch depths up to 50%, prism coupling lengths between 1 and 2 μm, and substrate radiation losses lower than 0.1 dB.

1) Measured (as described in chapters 7 and 8).
2) Inferred from references 4 and 5.

APPENDIX Weight coefficients for volume scattering loss

Unger derives a formula for the attenuation coefficient α which, after correction for a minus sign[1], has the following form (Ref. [2], equation 2.107)

$$\alpha\beta k_o d^2 = \{(vwD_u n_f \alpha_f - uwD_v n_s \alpha_s - uvD_w n_c \alpha_c)\, k_o^2 d^2 + uvw(D_f \alpha_f + D_s \alpha_s + D_c \alpha_c)\}$$

$$/\,(vwD_u - uwD_v - uvD_w), \qquad (23)$$

in which β is the propagation constant, k_o the vacuum wavenumber, d the film thickness, u, v, w, n_f, n_s, n_c, α_f, α_s, and α_c are the normalized transverse propagation constants, the refractive indices and the attenuation coefficients of the waveguiding film, the substrate, and the cover layer, respectively. The factors $D_{u,v,w,f,s,c}$ are the partial derivatives of the dispersion relation with respect to u, v, w, n_f, n_s, and n_c, and can be expressed as:

$$D_u = 1 + \frac{r_s v}{u^2 + r_s^2 v^2} + \frac{r_c w}{u^2 + r_c^2 w^2}, \qquad (24a)$$

$$D_v = -\frac{r_s u}{u^2 + r_s^2 v^2}, \qquad (24b)$$

$$D_w = -\frac{r_c u}{u^2 + r_c^2 w^2}, \qquad (24c)$$

with $r_s = (n_f/n_s)^{1/2q}$, $r_c = (n_f/nc)^{1/2q}$, $q=0$ for *TE*-modes and $q=2$ for *TM*-modes. The partial derivatives D_f, D_s, and D_c can be computed from:

1) The formula as given by Unger can lead to negative values of η_s and η_c. Inspection reveals that the second (bracketed) term in the numerator should have a positive sign.

$$D_f = q\,(vD_v + wD_w)/n_f \tag{24d}$$

$$D_s = -q\,D_v/n_s, \tag{24e}$$

$$D_c = -q\,D_w/n_c, \tag{24f}$$

If formula 3.23 is written in the form of equation 3.19 then we find for the weight coefficients η the following expressions:

$$\eta_f = vw\,(n_f D_u + \frac{uD_f}{k_o^2 d^2})/D, \tag{25a}$$

$$\eta_s = -uw\,(n_s D_v - \frac{vD_f}{k_o^2 d^2})/D, \tag{25b}$$

$$\eta_c = -uv\,(n_c D_w + \frac{uD_f}{k_o^2 d^2})/D, \tag{25c}$$

with

$$D = \frac{\beta}{k}(vwD_u - uwD_v - uvD_w). \tag{25d}$$

Substitution of equations 3.25a,b,c into equation 3.19 yields an expression for the contribution of the different layers to the attenuation of a mode.

References

3.1 H.Kogelnik, "Theory of dielectric waveguides", in: T. Tamir, Topics in Applied Physics: Integrated Optics (Berlin: Springer Verlag, 1979), pp.13-81.

3.2 H.-G. Unger, Planar optical waveguides and fibres (Oxford: Clarendon Press, 1977).

3.3 J. Chillwell and I. Hodgkinson, "Thin-films field-transfer matrix theory of planar multilayer waveguides and reflections from prism-loaded waveguides", J. Opt. Soc. Am. A, 1984, Vol. 1, pp. 742-753.

3.4 I.H. Malitson, "Refraction and dispersion of synthetic sapphire", J. Opt. Soc. Am., Vol. 52, pp. 1377-1397, 1962.

3.5 American Institute of Physics Handbook (New York: McGraw Hill Book Company, 1982).

3.6 R.M. Knox and P.P. Toulios, "Integrated circuits for the millimeter through optical frequency range", Procs. MRI Symp. on Submillimeter Waves, ed. J. Fox (Brooklyn NY: Polytechnic Press,1970), pp. 497-516.

3.7 P.A. Kirkby and G.B.H. Thompson, "Channeled substrate buried heterostructure GaAs-(GaAl)As injection lasers", Journal of Applied Physics, Vol. 47, pp 4578-4589, 1976.

3.8 J. Buus, "The effective index method and its application to semiconductor lasers", IEEE Journal of Quantum Electronics, Vol. QE-18, pp. 1083-1089, 1982.

3.9 J. Buus, "Application of the effective index method to nonplanar structures", IEEE Journal of Quantum Electronics, Vol. QE-20, pp. 1106-1109, 1984.

3.10 N.H.G. Baken, Nomenclature of guided wave modes in integrated- and fibre optics, Report 1 RNL/89, PTT Research, Neher Laboratories, Leidschendam, the Netherlands, April 1989.

3.11 N.H.G. Baken, private communication.

3.12 Schott catalogue, Optisches Glas, SF13.

3.13 F.H. Groen, private communication

3.14 B. Chapman, Glow discharge processes, sputtering and plasma etching (New York: Wiley, 1980).

3.15 R.J. Deri, E. Kapon, and L.M. Schiavione, "Scattering in low-loss GaAs/AlGaAs rib waveguides", Appl. Phys. Lett., Vol. 51, pp. 789-791, 1987.

3.16 R.J. Deri, R.J. Hawkins, and E. Kapon, "Rib profile effects on scattering in semiconductor optical waveguides", Appl. Phys. Lett., Vol. 53, pp. 1483-1485, 1988.

3.17 P.K. Tien, "Light waves in thin films and integrated optics", Applied Optics, Vol. 10, pp. 2395-2413, 1971.

3.18 D. Marcuse, "Mode conversion caused by surface imperfections of a dielectric slab waveguide", Bell. Syst. Tech. J., Vol. 48, pp. 3187-3216, 1969.

Chapter 4

Waveguide bends

In this chapter a brief introduction is given on the properties of modes in bent waveguides and the optimal design of junctions between straight and bent waveguides. Step-index waveguide bends are characterized by their bending radius, width and index contrast. Other relevant parameters are the width of the straight waveguides, to which they are connected, and the offset between straight and curved waveguides. A normalized approach is developed for the analysis of bends and junctions between straight and bent waveguides. With this approach an optimal bend design is possible from a set of normalized graphs.

4.1 Introduction

Waveguide bends are used both for interconnecting different components on one chip and as a component or part of a component, for example in ring resonators. The quality of waveguide bends is primarily determined by their losses and dimensions. Low-loss waveguide bends with small bending radii are indispensable to circuit miniaturization. Circuit miniaturization is a prerequisite for large-scale application of integrated optical circuits.

The first theoretical paper on bends in optical dielectric guides was published in 1969 by Marcatili [1]. Since then a large number of methods have been developed to analyze propagation through waveguide bends. A powerful technique is the conformal transformation method developed by Heiblum and Harris [2] in which the curved waveguide is translated into an equivalent straight one with a transformed index profile. A suitable method for solving the transformed problem is the Transfer-Matrix Method [3]. This method, which is well-known in optics, was applied to the conformally transformed index profile of curved waveguides by Thyagarajan et al. [4]. Results shown in the present chapter are computed with this method, as programmed by Pennings [5].

The first article concerning an experimentally demonstrated waveguide bend was published by Goell and Standley [6] in 1969. In the following decade few experiments on curved waveguides were reported [7-9]. A considerable number of papers appeared concerning waveguide bends in $LiNbO_3$, the first one being published by Ramaswamy and Divino [10] in 1981. These bends, however, have large bending radii due to the low index contrast in $LiNbO_3$ waveguides (lowest radius reported is 5.5 mm, by Korotky et al. [11]).

Short bending radii were first reported in 1983 by Austin, who measured 1 dB/90° for a bending radius of 300 μm in GaAs/AlGaAs rib guides [12,13]. Recently a number of publications on very short bending radii have appeared [14-17]. The best result reported is 0.6 dB/90° for 250 μm bending radius in

GaAs/AlGaAs by Rolland et al. [17]. Singh et al. [16] reported 1.1 dB/90° for 200 μm bending radius in GaInAsP/InP.

In Delft, research on waveguide bends was initiated in 1984 with an investigation into the applicability of bends composed of coherently coupled straight waveguide sections as proposed by Taylor [18,19]. Experimental work was begun by R.V. Funke [20] and continued by G.H. Manhoudt [21]. They concluded that submillimeter radii were difficult to realize employing this approach. Parallel research of Manhoudt [21] on smooth waveguide bends yielded very promising experimental results for bending radii as small as 175 μm. Since 1986 this work was continued by E.C.M. Pennings. His work [22-29] yielded the shortest low-loss bends so far reported (0.6 dB/90° for Al_2O_3 waveguides with R=50 μm at λ=633 nm, 0.5 dB/90° for InGaAsP/InP waveguides with R=150 μm at λ=1.52 μm).

Optimal bend design requires the determination of five parameters for a bend with a given index contrast and bending radius: the radiation loss, the optimal widths of the bend and the straight waveguides to which it has to be connected, the optimal offset between these waveguides, and the ultimate coupling loss. An impediment to the design of optimal bends is the rather complicated analysis which is required to determine these parameters. Pennings [5] computed normalized solutions for three of the five parameters mentioned which greatly reduce the required analysis. In this chapter the normalized approach will be extended so as to provide normalized solutions to all relevant design parameters.

4.2 Guided modes in bent waveguides

In order to remain guided, the phase fronts of a mode in a curved waveguide must rotate as depicted in figure 4.1. This causes the spatial period, measured in the ϕ-direction along an arc with constant radius to become dependent on the radius ρ. The wavelength will increase linearly with ρ and the propagation constant β will, therefore, be inversely proportional to ρ. There will thus exist a radius ρ_c for which the local effective index β/k will become smaller than the refractive index of the surrounding medium. In the region between the waveguide and this point the amplitude of the guided mode

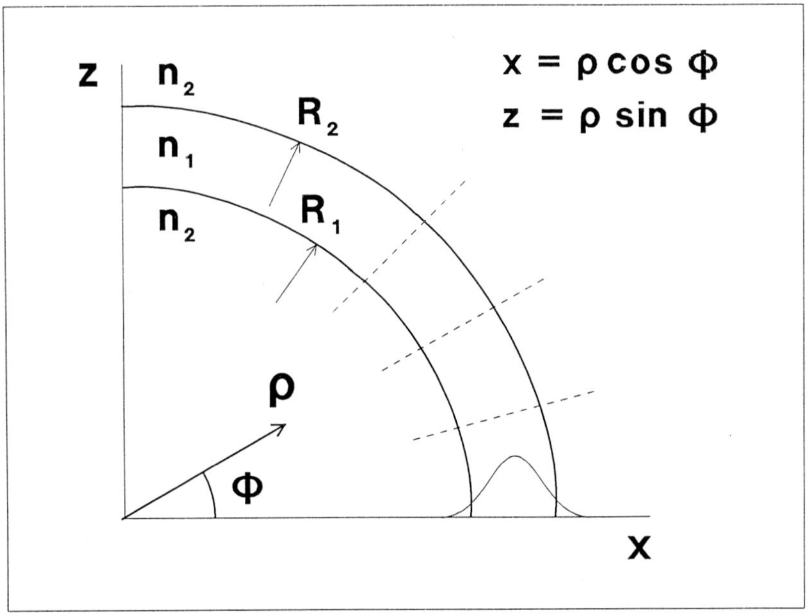

Figure 4.1 Cylindrical coordinate system for the analysis of wave propagation in a curved waveguide.

will decay exponentially; for greater values of ρ it will become oscillatory, which means that the mode will experience radiation loss.

From this qualitative analysis it can be seen that radiation loss is inherent to waveguide bending. For short bending radii the ρ-dependence of β will be large and the point at which the mode profile becomes oscillatory will be close to the waveguide, so that considerable radiation losses may be expected.

A more quantitative understanding of the wave propagation phenomena in curved waveguides can be obtained by translating the "curved" problem into an equivalent "straight" one. In a curved waveguide we expect the wave propagation to be described by an angular propagation constant β_ϕ:

$$U(\rho,\phi) = U(\rho)\, e^{-j\beta_\phi \phi} . \qquad (1)$$

We can assign a reference propagation constant β_t to this cylindrical mode as the wavenumber measured along an arc with (arbitrary) reference radius R_t:

$$\beta_t = \frac{d\Phi_t}{ds_t} = \frac{\beta_\phi\, d\phi}{R_t\, d\phi} = \frac{\beta_\phi}{R_t}, \qquad (2)$$

in which Φ_t and s_t are the phase of the mode and the spatial coordinate measured along the arc with radius R_t, respectively. The propagation constant $\beta(\rho)$ being inversely proportional to ρ, it follows that:

$$\beta(\rho) = \beta_t R_t / \rho. \tag{3}$$

In a straight waveguide the propagation constant β and the mode profile follow from the Helmholtz equation together with the boundary conditions at discontinuous interfaces and at infinity:

$$\frac{\partial^2}{\partial y^2} U + k_y^2(y) U = 0, \tag{4}$$

in which $k_y(y)$ is the lateral propagation constant:

$$k_y(y) = \{n^2(y)k_o^2 - \beta^2\}^{1/2}. \tag{5}$$

The choice of the principal interval of the root k_y is determined by the type of the mode which we are searching for; guided or leaky[1]. Substitution of the expression for $\beta(\rho)$ into equation 4.5 gives:

$$k_\rho(\rho) = \{n^2(\rho)k_o^2 - \beta_t^2 \frac{R_t^2}{\rho^2}\}^{1/2} \simeq \{n^2(\rho)\frac{\rho^2}{R_t^2} k_o^2 - \beta_t^2\}^{1/2} \tag{6}$$

The latter approximation is valid if $R_t / \rho \simeq 1$ in the vicinity of the waveguide; it can be expressed as:

$$k_\rho(\rho) \simeq \{n_t^2(\rho)k_o^2 - \beta_t^2\}^{1/2}, \tag{7}$$

1) For guided modes we choose k_y such that $Im(k_y)<0$, for leaky modes we choose $Im(k_y)>0$ in the semispace into which the mode is radiating. For more details see Blok et al. [30].

in which

$$n_t(\rho) = n(\rho)\frac{\rho}{R_t}. \tag{8}$$

Obviously the mode profile $U(\rho)$ in a curved waveguide with index profile $n(\rho)$ can be computed as the mode profile of an equivalent straight waveguide with a transformed index profile $n_t(\rho)$. The transformed index profile and the corresponding mode profile are illustrated in figure 4.2. The angular propagation and attenuation constants β_ϕ and α_ϕ (defined according to equations 3.14-16) follow from β_t and α_t through equation 4.2:

$$\beta_\phi = \beta_t R_t, \tag{9a}$$

$$\alpha_\phi = \alpha_t R_t. \tag{9b}$$

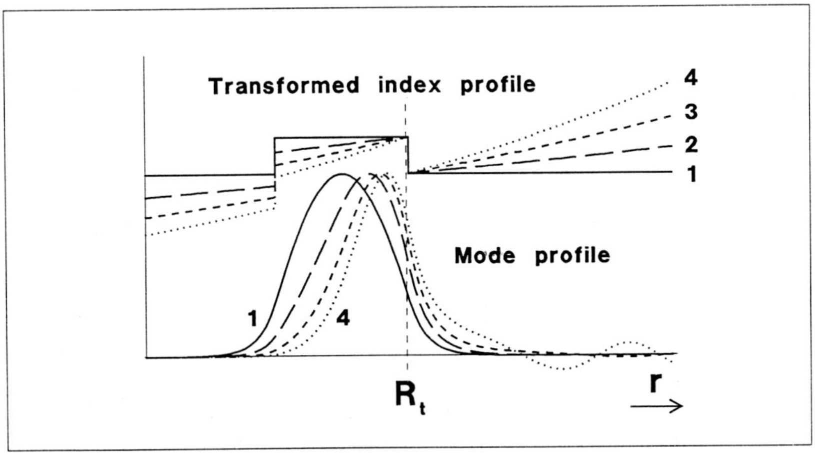

Figure 4.2 The transformed index profile and the corresponding mode profile for a straight waveguide (1) and curved waveguides with decreasing bending radius (2-4).

The choice of the reference radius R_t is arbitrary; a greater value of R_t leads to a lower value of β_t through equations 4.7 and 4.8. This is compensated, however, by multiplication with R_t in equation 4.9. A practical choice would be to set it equal to the bending radius at the outer edge of the waveguide.

Comparison with the approach of Heiblum and Harris [2] (see appendix 3A) reveals that the transformation applied above is the linear approximation of the one applied by these authors. This exact transformation has been employed throughout this chapter for numerical computations. The radiation loss follows from the computed value of β_t as:

$$\alpha_\phi = - Im\,(\beta_t)\,R_t, \qquad rad^{-1} \tag{10}$$

or

$$A_\phi = 20\ log_{10}\,(\alpha_\phi \pi/2). \qquad dB/90° \tag{11}$$

A more detailed description of the computational method[1] is given by Thyagarajan et al. [4] and Pennings [5].

1) A complication follows from the fact that the boundary conditions are invariant under the transformation, as may be easily verified. For TM-polarized modes (i.e. modes for which the H-field only has a radial component) the second boundary condition requires that $(1/n^2)\cdot\partial/\partial\rho$ be continuous through a step in the index profile. Because the boundary condition applies to the original index profile, a modification of the transfer-matrix method is necessary in order that the transfer through a layer is inferred from the transformed profile, whereas the transfer through an interface is inferred from the original profile. A thus-modified method was programmed by Pennings [5] and used by the present author for numerical computations. The problem does not occur for TE-modes. The error in employing a non-modified method will be small for low index contrasts.

4.3 Optimal bend design

From figure 4.2 it can be seen that the mode profile will shift to the outer edge of the waveguide if the bending radius is decreased. If a sufficiently small bending radius is chosen the field strength at the inner edge vanishes and the mode will be fully guided by the outer edge (curves 3 and 4) so that the location of the inner edge becomes irrelevant. Such a mode is called a Whispering Gallery Mode (WG-mode) after Lord Rayleigh [31] who explained this phenomenon in relation to the propagation of sound waves along a curved gallery.

Because the field profile of a whispering-gallery mode will differ from that of a straight waveguide, coupling loss will occur at the junctions between curved and straight waveguides. It can be minimized by matching the two mode profiles as good as possible. This can be achieved through a proper choice of the width and the location of the straight waveguide relative to the bent one, as illustrated in figure 4.3, such that the overlap between the straight and the bent waveguide mode is optimal.

The application of an offset between the straight and the curved waveguide in order to reduce transition loss was proposed by Neumann [32] in 1982. Four years later Manhoudt [21] recognized that not only the relative alignment, but also the width of the straight waveguide should be optimized.

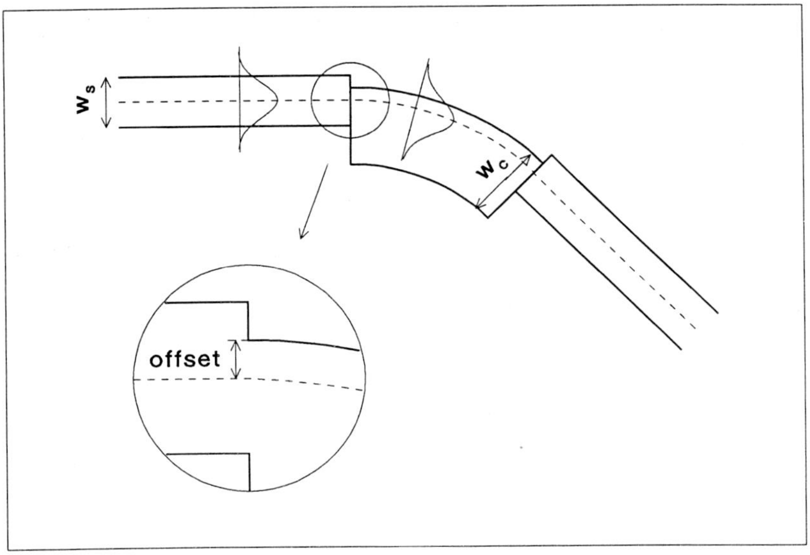

Figure 4.3 *A curved and a straight waveguide section, which are optimally dimensioned and aligned for small transition loss.*

Sheem and Whinnery [33] had demonstrated in 1974 that Whispering Gallery Modes can be applied to integrated optical circuits. Pennings [5] showed that the lowest total bending loss is obtained by employing curved waveguides which operate in the whispering-gallery-mode regime. He studied the problem of optimal bend design in depth, and was the first to obtain experimental losses close to the theoretical limits.

The optimization of the bending loss, as described above, is straightforward, but too laborious to be performed without dedicated software. The analysis can be simplified, however, by a proper normalization of the problem.

4.4 Normalized approach to optimal bend design

A normalized approach to the analysis of bent waveguides was applied by Marcatili [1] and further elaborated by Pennings [5]. In this section the approach will be extended so as to provide normalized solutions to all relevant parameters for optimal design of waveguide bends including the junctions with straight waveguides.

We firstly normalize all dimensions on the wavelength λ_2 in the background medium:

$$\chi = \frac{y}{\lambda_2}. \tag{12}$$

In terms of the normalized coordinate χ and the relative index contrast profile:

$$\Delta(\chi) = \frac{n(\chi) - n_2}{n_2}, \tag{13}$$

the Helmholtz equation (4.4) transforms into:

$$\frac{\partial^2}{\partial \chi^2} U(\chi) + 4\pi^2 [\{1+\Delta(\chi)\}^2 - \{1+\Delta_e\}^2] U(\chi) = 0, \tag{14}$$

in which Δ_e is the effective relative index contrast, which is related to the effective index N through:

$$\Delta_e = \frac{N - n_2}{n_2}. \tag{15}$$

For $\Delta(\chi) \ll 1$ equation 4.14 reduces to:

$$\frac{\partial^2}{\partial \chi^2} U(\chi) + 8\pi^2\{\Delta(\chi) - \Delta_e\} U(\chi) \simeq 0. \tag{16}$$

If we introduce a new variable ξ:

$$\xi = \chi / a, \tag{17}$$

equation 3.16 transforms into $(\frac{\partial^2}{\partial \xi^2} = \frac{1}{a^2} \frac{\partial^2}{\partial \chi^2})$:

$$\frac{\partial^2}{\partial \xi^2} V(\xi) + 8\pi^2\{\overline{\Delta}(\xi) - \overline{\Delta}_e\} V(\xi) = 0. \tag{18}$$

in which

$$\overline{\Delta}(\xi) = a^2\, \Delta(a\xi), \tag{19}$$

$$\overline{\Delta}_e = a^2\, \Delta_e, \tag{20}$$

$$V(\xi) = U(a\xi). \tag{21}$$

From these equations it may be seen that if $\{U(\chi), \Delta_e\}$ is a solution of equation 4.16 for the index profile $\Delta(\chi)$, then $\{U(a\chi), a^2\Delta_e\}$ is a solution for the index profile $a^2\Delta(a\chi)$, i.e. if the relative contrast profile is compressed by a factor a and its height is multiplied by a factor a^2, as shown in figure 4.4,

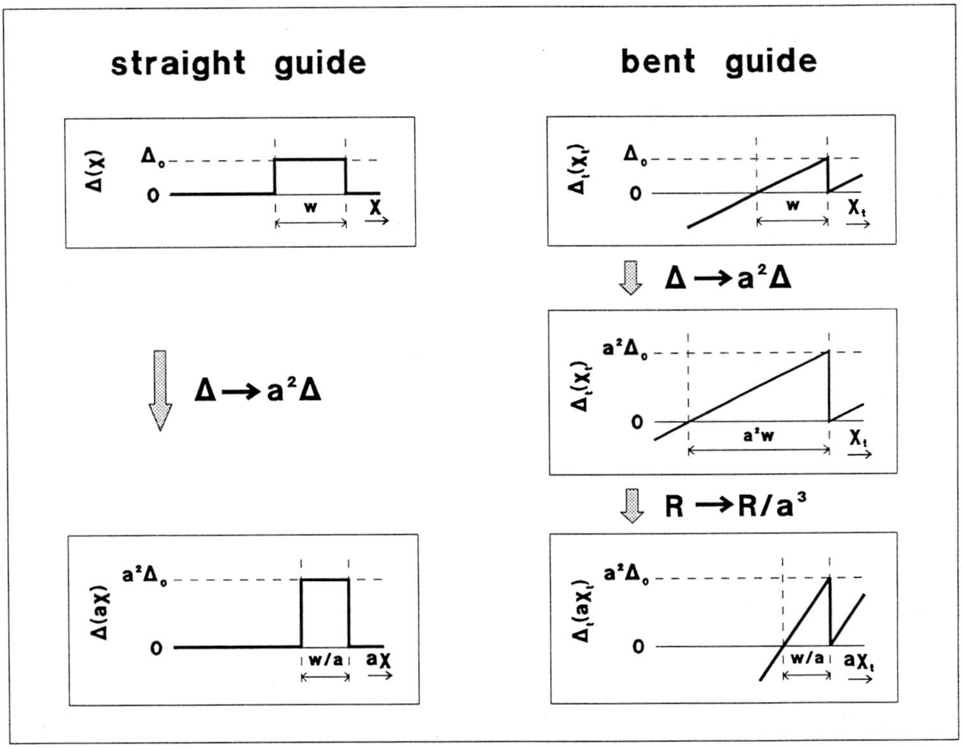

Figure 4.4 Transformation of the index profile of a straight and a bent waveguide, such that the shape of the mode profile is not changed.

then the mode profile is compressed by a factor a, but otherwise retains the same shape.

If we apply this transformation to the (transformed) index profile of a bend with a relative index contrast Δ_o and a normalized radius ρ, we see from figure 4.4 that the width of the transformed index profile is increased by a factor of a^2 (the figure illustrates the transformation for $a>1$; the formula's also apply to $a<1$). If we reduce the bending radius by a factor a^3, the resulting width of the transformed profile will become smaller than the original by a factor a. We may thus conclude that the mode profile for a waveguide with index contrast $\Delta=a^2\Delta_o$ and bending radius R can be derived from that of a bend with contrast Δ_o and radius a^3R through compression by a factor a.

The propagation constant β and the attenuation coefficient α are related to the effective index contrast Δ_e of the mode as (see equations 3.14-16 and 4.15):

$$\beta = k_o n_2\{1+Re(\Delta_e)\}, \tag{22a}$$

$$\alpha = k_o n_2\{Im(\Delta_e)\}. \tag{22b}$$

From equation 4.22b, in combination with equation 4.20 which also holds for complex values of Δ_e, it follows that the attenuation coefficient α_t transforms according to:

$$\bar{\alpha}_t = a^2\alpha_t. \tag{23a}$$

The angular attenuation coefficient α_ϕ follows from the transformed constant α_t through multiplication by R_t (equation 4.9b). Because R_t is reduced by a factor of a^3 the total effect of the transformation on α_ϕ is a reduction by a factor of a:

$$\bar{\alpha}_\phi = \alpha_\phi/\alpha. \tag{23b}$$

Obviously, both the mode profile in bent and straight waveguides, and the radiation loss in bent waveguides transform in a very simple manner under the transformation of equation 4.17. This is an important result. It means that, if we compute the radiation loss or the coupling loss at the junction between a straight and a bent waveguide for a bending radius R and a contrast Δ_o, then the properties for other contrasts Δ can be directly inferred, as long as both Δ and Δ_o are small:

$$\Delta, \Delta_o \ll 1. \tag{24}$$

It should be stressed that the above normalization only applies to small index contrasts, and that the bent waveguides should operate in the whispering-gallery mode regime. If the inner waveguide edge affects the mode width, the accuracy of the transformation expressions 4.19-21 and 4.23 will be reduced.

4.5 Numerical results and empirical corrections

On the basis of the normalization described in the previous section, the analysis of a waveguide with arbitrary contrast Δ and bending radius R can be reduced to the analysis of a waveguide with a normalized index contrast Δ_o and a normalized bending radius $a^3(R/\lambda_2)$ with $a=(\Delta/\Delta_o)^{1/2}$, as long as $\Delta \ll 1$. For our analysis we chose $\Delta_o=0.01$. To design a waveguide bend with low loss and optimal junctions to the straight waveguides, the following five quantities have to be determined:

- the angular radiation loss,
- the minimal width of the curved waveguide.
- the optimal width of the straight waveguide
- the optimal offset between the curved and the straight waveguide
- the corresponding coupling loss

Figures 4.5a-d show the results of the analysis as a function of the normalized radius. It was empirically determined that, for the range of radii depicted in figures 4.5a-d, a minimal width w_c of the curved waveguide:

$$w_c = 1.5\, w_s\,, \qquad (25)$$

is sufficient to obtain whispering-gallery-mode operation for the fundamental mode. For small contrasts the normalized solutions apply to *TE*-polarized as well as *TM*-polarized modes. For the maximal relative contrast analyzed in the present chapter $(\Delta=0.16)$ the *TM*-radiation loss was found to be greater by 30% then the *TE*-polarized loss. The normalized optimal offset for *TM*-polarization was found to be smaller than the *TE*-polarized value by approximately 0.2 μm. The difference in optimal waveguide width is within 1%. Differences between the normalized solutions for both polarizations are thus negligible for most practical purposes. It is stressed that in three-dimensional waveguides the polarization-dependence may be greater because the effective indices of the transverse slab modes which form the starting point for the lateral computations may differ considerably. This effect can be analyzed, however, using the normalized approach.

The figures are employed as follows. The normalized radius R_n is computed according to:

$$R_n = a^3 R/\lambda_2 * 1.137^{\Delta-\Delta_o} \qquad a=(\Delta/\Delta_o)^{½} \qquad (26)$$

in which R and Δ are the actual radius and refractive index contrast of the waveguide, and Δ_o=*0.01* is the value of Δ for which the graphs were computed. The origin of the correction factor $1.137^{\Delta-\Delta_o}$ will be discussed in the sequel. The required properties can then be read from the relevant graph. The coupling loss is independent of the normalization and can be read directly. The other properties are determined on the basis of the normalized values, as read from the figure, through division by the product of a $(=\Delta/\Delta_o)^{½}$ and the correction factor as listed along the vertical axis. A polynomial description of the curves is provided in Appendix 4B at the end of this section. Details on the numerical analysis are given in Appendix 4C.

To analyze the accuracy of the normalization, we have computed the radiation loss, the normalized width and the normalized offset for a series of contrasts, ranging from 0.0025 to 0.16. The relative error has been determined by dividing these results by those computed using the normalized solutions. Figures 4.6a,c,e show the result. From figure 4.6a we see that the relative error in A_ϕ is linear in R and exponential in Δ. Because the dependence of the logarithm of A_ϕ on R is approximately linear, the error can be compensated with a correction term of the form $c^{\Delta-\Delta_o}$. Calculation yields c=*1.137* as a good fit. Figure 4 .6b shows the resulting error after correction. Its magnitude appears to be linear with Δ, from which we conclude that the normalization error will be within 20% for contrasts up to 0.2.

The errors in the normalized offset and width (of the straight waveguide) appear to be independent of R and linear in Δ. This again suggests a correction factor of the form $c^{\Delta-\Delta_o}$. For the offset a good fit is found with c=*2*, for the optimal width with c=*1.75*. Figures 4,6d and 4.6f show the relative error after correction, which appears to be within 6% for the offset (within 2% for R_n>*1000*) and within 2% for the width. The markers in figure 4.5a show the effects of the residual errors for the least and the greatest contrast (0.0025 and 0.16 respectively), at the extreme ends of the computation range. From these data it is evident that the errors will be negligible for almost all practical purposes.

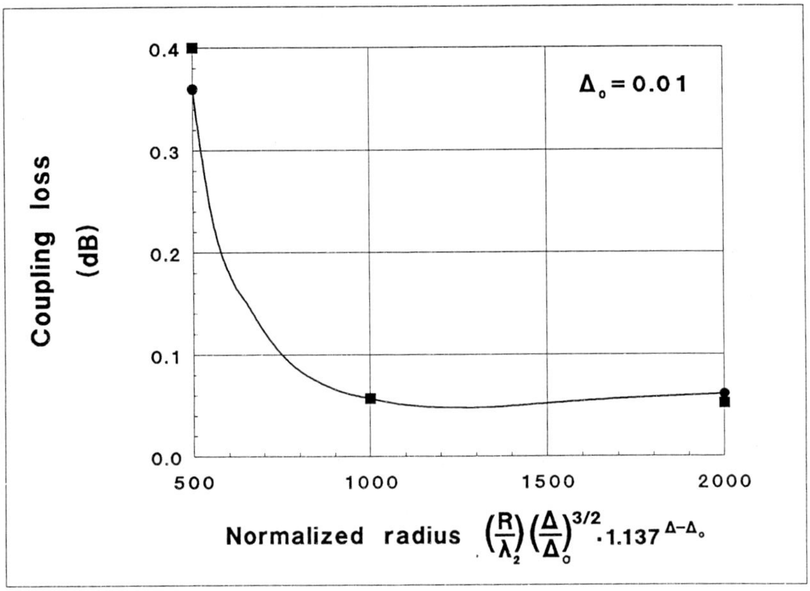

Figure 4.5a *Normalized transition loss for a single junction.*

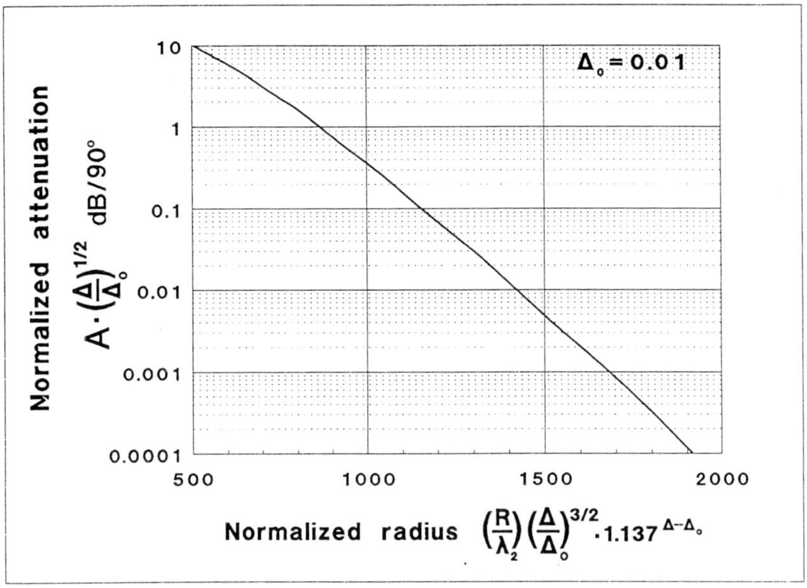

Figure 4.5b *Normalized radiation loss.*

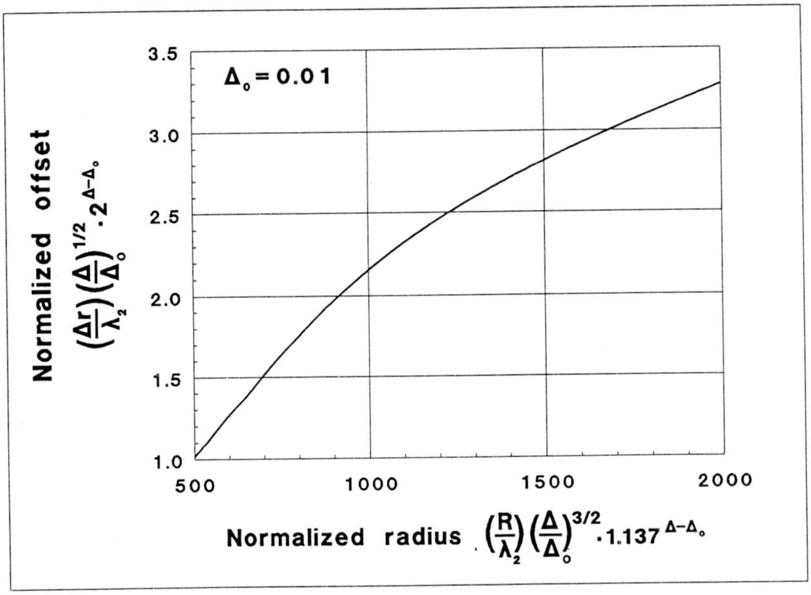

Figure 4.5c *Normalized optimal offset between the outer edge of the curved waveguide and the center of the straight one.*

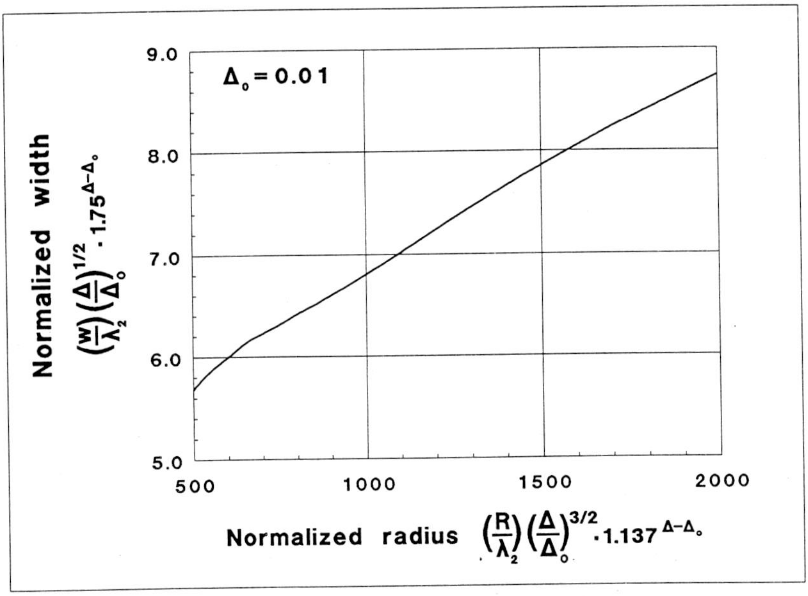

Figure 4.5d *Normalized optimal width of the straight waveguide.*

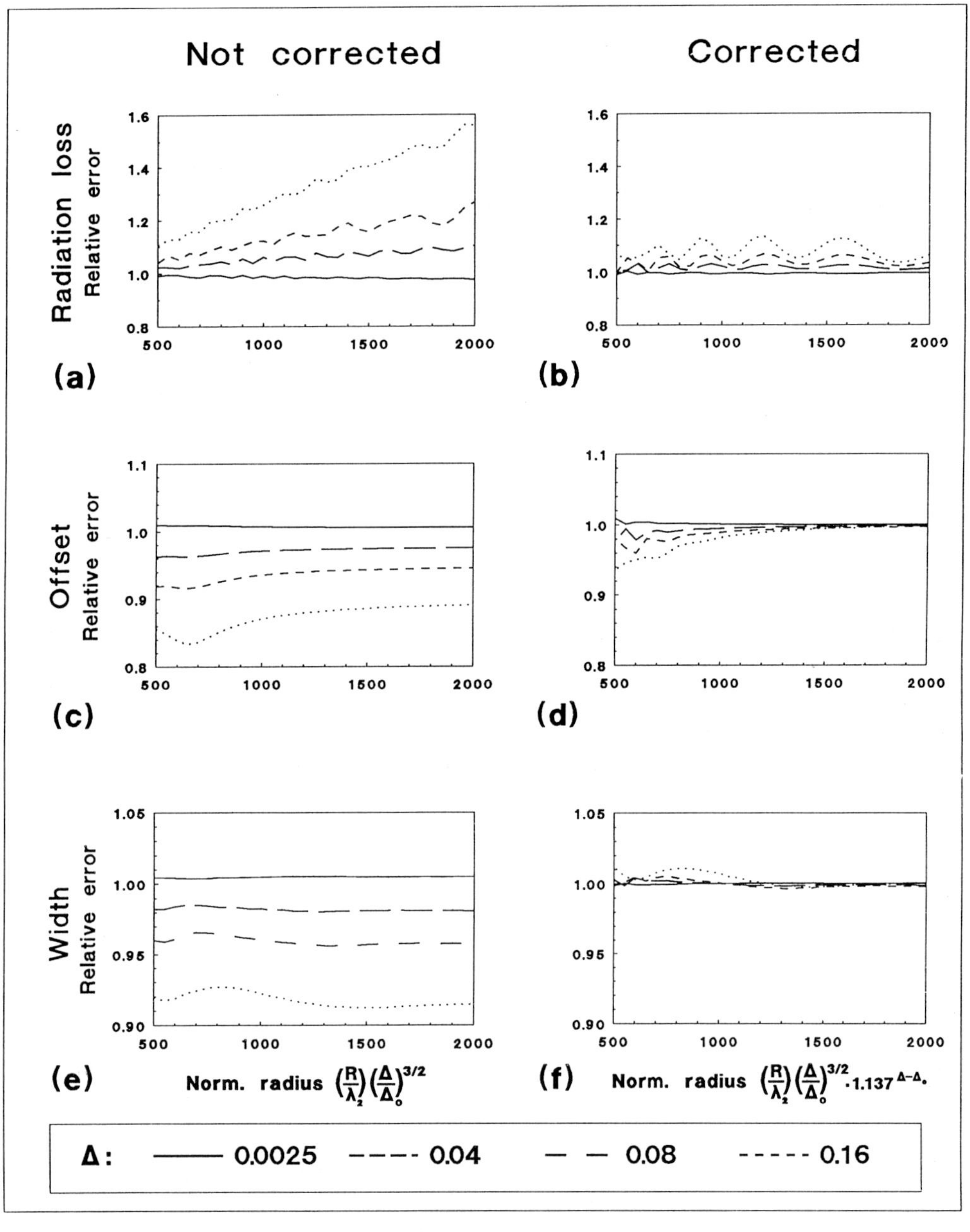

*Figure 4.6 Relative errors in the normalized solutions for (**a**) the radiation loss, (**c**) the optimal offset, and (**e**) the optimal width, for different index contrast, without (**a,c,e**) and with empirical corrections as described in the text (**b,d,f**).*

4.6 Discussion and conclusions

Employing the normalized graphs of figure 4.5 or the regression formulae of Appendix 4B optimal bend design can be performed with a pocket calculator for a broad variety of planar optical waveguides with low or medium optical contrast. Two different design strategies will be briefly discussed.

If the lowest possible loss is required, a normalized radius should be selected for which the sum of the radiation loss (over the relevant sector angle) plus twice the coupling loss is minimal. Except for very low contrasts the total loss will be dominated by the coupling loss and a normalized radius between 1000 and 1500 will be optimal, corresponding to a normalized radiation loss between 0.4 and 0.005 dB/90°. The corresponding optimal widths and the offset between the straight and the curved waveguides follow from figures 4.5c,d and equation 4.19.

If the choice of the straight-waveguide width is not free the radius of the bend has to be chosen such that the mode width matches that of the straight waveguide. This is done by reading the normalized radius corresponding to the prescribed (normalized) width of the straight waveguides from figure 4.5d. The other parameters are fixed by this choice, and follow from the graphs.

The normalized approach as outlined above is particularly suited for computer-aided design and simulation of planar optical circuits. We confined ourselves to the excitation of the fundamental modes which cover most of the practical applications. Radiation loss and coupling efficiencies for higher-order modes can be normalized equally well. For the method to be applicable index contrasts should be low and the bent waveguide should be sufficiently wide (whispering-gallery-mode regime).

APPENDIX 4A Guided waves in a cylindrical coordinate system

In order to arrive at a quantitative understanding of wave propagation in a curved waveguide we will have to find solutions of the Helmholtz equation in the cylindrical coordinate system of figure 4.1.

Transformation of the Helmholtz-equation and separation of variables according to $U(\rho,\phi) = R(\rho)\,\Phi(\phi)$ brings us to the following expressions:

$$\left\{\frac{\partial^2}{\partial\rho^2} + \frac{1}{\rho}\frac{\partial}{\partial\rho} + k_o^2 N^2(\rho) - \frac{\beta_\phi^2}{\rho^2}\right\} R(\rho) = 0, \tag{27}$$

$$\left\{\frac{\partial^2}{\partial\phi^2} + \beta_\phi^2\right\} \Phi(\phi) = 0, \tag{28}$$

in which β_ϕ^2 is the separation constant. The latter equation has general solutions of the form:

$$U(\phi) = C\, e^{\pm j\beta_\phi \phi}. \tag{29}$$

The solutions resemble those for straight waveguides, the phase fronts now coinciding with planes of constant ϕ instead of constant z. De constant β_ϕ can be looked upon as an angular propagation constant with dimension rad^{-1}.

The admitted values of β_ϕ and the corresponding mode profile $U_\beta(\rho)$ follow by solving equation 4.27. With the following transformation, which was proposed by Harris and Heiblum [5]:

$$\rho_t = R_t \, ln\,(\rho/R_t), \tag{30}$$

the equation for $U(\rho)$ is brought onto the form:

$$\left\{\frac{\partial^2}{\partial\rho_t^2} + \{k_o^2\, n_t^2\,(\rho_t) - \beta_t^2\}\right\} V(\rho_t) = 0, \tag{31}$$

in which

$$n_t(\rho_t) = n(\rho)\, e^{\rho/R_t}, \tag{32}$$

and

$$\beta_t \quad = \beta_\phi/R_t\,. \tag{33}$$

Equation 4.31 has the same shape as the Helmholtz equation for a straight waveguide with index profile $n_t(y)$. Linearization of the transformed index profile leads to the expression of formula 4.8. For numerical analysis purposes the exact transformation according to equation 4.32 is easily programmed. All results presented in this thesis are computed with the latter transformation.

APPENDIX 4B Empirical formulae for optimal bend design

The normalized curves of figures 4.5a-d are easily quantified with polynomial regression. The results are given below in terms of the real (i.e. not normalized) parameters, for $\Delta_o = 0.01$:

$$A_\phi = (100\,\Delta)^{-1/2}\, 10^{1.63 - 5.97 R'_n + 3.92 (R'_n)^2 - 0.82 (R'_n)^3}, \quad (dB/90°), \quad (33)$$

= radiation loss per 90° in dB.

$$w_s = \frac{\lambda_o}{n_2} (100\,\Delta)^{-1/2} \{4.65 + 2.45 R'_n - 0.18 (R'_n)^2\}\, 1.75^{0.01-\Delta}, \quad (34)$$

= optimal width of the straight waveguide.

$$w_c = 1.5\, w_s, \quad (35)$$

= minimal width of the curved waveguide.

$$\Delta r = \frac{\lambda_o}{n_2} (100\,\Delta)^{-1/2} \{-0.9 + 4.7 R'_n - 2.0 (R'_n)^2 + 0.35 (R'_n)^3\}\, 2^{0.01-\Delta}, \quad (36)$$

= optimal offset between the outer edge of the curved waveguide and the center of the straight one.

$$\eta = 10^{1.63 - 5.97 R'_n + 3.92 (R'_n)^2 - 0.82 (R'_n)^3}, \quad (dB), \quad (37)$$

= coupling loss in dB at a (single) junction between a straight and a curved waveguide, optimized according to the above parameters.

$$R'_n = \frac{n_2 R}{\lambda_o} (100\,\Delta)^{3/2}\, 1.137^{\Delta - 0.01} / 1000, \quad (38)$$

= normalized bending radius (at the outer edge). Note the factor 1000 in the denominator, which is included to avoid repetition of factors 0.001 in the regression formulae).

The above formulae apply in the range $0.5 < R'_n < 2$, i.e. $500 < R_n < 2000$.

APPENDIX 4C Computation of bend properties

The computation of the angular radiation loss is straightforward. For our computations we used a normalized waveguide width of 20, which guarantees whispering-gallery-mode operation throughout the normalized radius range (500-2000). The index profile was extended at the outer edge over a normalized distance of 20. In the whispering-gallery-mode regime no extension is required at the inner edge. For the transfer-matrix method a subdivision of the index profile into layers with unit normalized width proved adequate to arrive at stable outcomes. These parameters have also been used for the computation of the other parameters.

The optimal width of the straight waveguide was calculated as follows. Firstly the effective width of the whispering-gallery mode was computed according to the following formula (equation 3.9):

$$w_e = \frac{\int U^2(\rho)\, d\rho}{U^2_{max}}. \tag{39}$$

The integration was extended over the (normalized) range 0-25, which may introduce a small error for normalized radii close to 500. The effective mode width w_s for the straight waveguide was computed from:

$$w_e = w\,(1+2/v), \tag{40}$$

in which w is the waveguide width and v the normalized transverse attenuation coefficient, as defined in chapter 3. Next we determined the coupling efficiencies at a number of normalized radii *(500, 600, 850, 1000, 1250, 1500, 2000)* for a series of waveguide widths ranging from 4 to 10 in steps of 1 and determined the optimal widths by quadratic interpolation. It was then found that the following estimate:

$$w \simeq w_e + 1, \tag{41}$$

gives an error smaller than ± 0.2 over the range $5 < w < 9$.

For the computation of the normalized offset several approaches are possible. Pennings [5] defined the offset as the distance between the maximum of the WG-mode and the outer edge of the waveguide. If the mode profile becomes strongly asymmetric this approach will introduce an error which increases with decreasing bending radius. We therefore defined the offset as the distance between the outer edge and the first moment of the intensity distribution:

$$\Delta\rho = \frac{\int \rho U^2(\rho)\, d\rho}{U^2_{max}}, \tag{42}$$

in which the integration was extended over the range $0<\rho<25$, which proved to give a better coupling efficiency at very short radii. In the range of practical values $(\rho>1000)$ the differences are negligible.

The coupling efficiencies were computed by overlapping the (normalized) mode profiles according to the normalized parameters as determined from the regression formulae of Appendix 4B. The computation was done for the following normalized radii: *500, 550, 600, 650, 750, 850, 1000, 1250, 1500,* and *2000*. To analyze the accuracy of the normalization the coupling efficiency was computed without normalization for the contrasts 0.0025 and 0.16, at the radii corresponding to $\rho=500$ and $\rho=2000$. The results are indicated by the markers in figure 4.5a.

It was determined empirically that for relative index contrasts in the range 1.0025-1.16 and normalized radii in the range 500-2000 the width w_c of the curved waveguide had a negligible effect on the above parameters if it was at least 1.5 times the optimal width w_s of the straight one.

All computations were performed using the waveguide analysis program WGANAL (version 2.3) of Delft University, which was developed by the present author. The program is based on the Effective Index Method in combination with the Transfer Matrix Method or the Modified Transfer Matrix Method[1] for the solution of two-dimensional problems. The routines developed by E.C.M. Pennings were used for the implementation of the latter methods.

References

4.1 E.A.J. Marcatili, "Bends in optical dielectric guides", Bell System Technical Journal", Vol. 48, pp.2103-2132, 1969.

4.2 M. Heiblum and J.H. Harris, "Analysis of curved optical waveguides by conformal transformation", IEEE Journal of Quantum Electronics, Vol. QE-11, pp. 75-83, 1975. Correction: Vol. QE-12, p. 313, 1976.

4.3 J. Chilwell and I. Hodgkinson, "Thin-films field-transfer matrix theory of planar multilayer waveguides and reflection from prism-loaded waveguides", Journal of the Optical Society of America A, Vol. 1, pp.742-753, 1984.

4.4 K. Thyagarajan, M.R. Shenoy, and A.K. Ghatak, "Accurate numerical method for the calculation of bending loss in optical waveguides using a matrix approach", Optics Letters, Vol. 12, pp. 296-298, 1987.

4.5 E.C.M. Pennings, "Bends in Optical ridge waveguides", Ph.D Thesis, ISBN 90-9003413-7, Delft University of Technology, Delft, 1990.

4.6 J.E. Goell and R.D. Standley, "Sputtered glass waveguides for integrated optical circuits", Bell Systems Technical Journal, Vol. 48, pp.3445-3448, 1969.

4.7 J.E. Goell, "Electron-resist fabrication of bends and couplers for integrated optical circuits", Applied Optics, Vol. 12, pp. 729-736, 1973.

---

1) See the note on page 98.

4.8 J.E. Goell, "Rib waveguide for integrated optical circuits", Applied Optics, Vol. 12, pp. 2797-2798, 1973.

4.9 H. Furuta, H. Noda, and A. Ihaya, "Novel optical waveguide for integrated optics", Applied Optics, Vol. 13, pp.322-326, 1974.

4.10 V. Ramaswamy and M.D. Divino, "Low loss bends for integrated optics", Procs. Conf. on Lasers and Electrooptics, Washington DC, 1981, paper THP1.

4.11 S.K. Korotky, E.A.J. Marcatili, J.J. Veselka, and R.H. Bosworth, "Greatly reduced losses for small-radius bends in Ti:$LiNbO_3$ waveguides", Applied Physics Letters, Vol. 48, pp.92-94, 1986.

4.12 M.W. Austin, "GaAs/AlGaAs curved rib waveguides", Journal of Quantum Electronics, Vol. QE-18, pp. 795-800, 1983.

4.13 M.W. Austin and P.G. Flavin, Small-radii curved rib waveguides in GaAs/AlGaAs using electron-beam lithography", Journal of Lightwave Technology, Vol. 1, pp. 236240., 1983.

4.14 H. Takeuchi and K. Oe, "Very low loss GaAs-AlGaAs miniature bending waveguide with curvature radii less than 1 mm", Applied Physics Letters, Vol. 54, pp.87-89, 1989.

4.15 H. Takeuchi and K. Oe, 'Low-loss single-mode GaAs/AlGaAs miniature optical waveguides with straight and bending structures", Journal of Lightwave Technology, Vol. 7, pp. 1044-1054, 1989.

4.16 J. Singh, I. Henning, M. Harlow, and S. Cole, "Single-mode low-loss buried optical waveguide bends in GaInAsP/InP fabricated by dry etching", Electronics Letters, Vol. 25, pp. 899-900, 1989.

4.17 C. Rolland, G. Mak, K.E. Fox, D.M. Adams, A.J. Springthorpe, D. Yevick, and B. Hermansson, "Analysis of strongly guiding rib waveguide S-bends: theory and experiments", Electronics Letters, Vol. 25, pp.1256-1257, 1989.

4.18 H.F. Taylor, "Power loss at directional change in dielectric waveguides", Applied Optics, Vol. 13, pp. 642-647, 1974.

4.19 H.F. Taylor, "Losses at corner bends in dielectric waveguides", Applied Optics, Vol. 16, pp.711-716, 1977.

4.20 R.V. Funke, "Demping in bochten in vlakke optische golfgeleiders" (Attenuation in planar optical waveguide bends), Masters Thesis, Delft University of Technology, Dept. of Electrical Engineering, Microwave Laboratory, Delft, 1985.

4.21 G.H. Manhoudt, "Bochten in planaire optische golfgeleiders, een theoretische en experimenteel onderzoek" (Planar optical waveguide bends, a theoretical and experimental investigation), Masters Thesis, Delft University of Technology, Dept. of Electrical Engineering, Microwave Laboratory, Delft, 1986.

4.22 E.C.M. Pennings, G.H. Manhoudt and M.K. Smit, "Low-loss bends in planar optical ridge waveguides", Electronics Letters, Vol. 24, pp. 998-999, 1988.

4.23 E. van der Drift, B.A.C. Rousseeuw, J. Romijn, F.H. Groen and E.C.M. Pennings, "High resolution patterning of aluminium oxide for optical devices", Microcircuit Engineering, ME88, Vol. 7, Procs. of the International Conference on Microlithography, Vienna, Austria, Sptember 20-22, 1988, Amsterdam: North-Holland (1988).

4.24 E.C.M. Pennings, "Small-radii curved waveguides for integrated optoelectronics", abstract. SIOE'89, University of Wales College of Cardiff, UK, 20/21 March 1989.

4.25 E.C.M. Pennings, J. van Schoonhoven, J.W.M. van Uffelen, and M.K. Smit, "Reduced bending and scattering losses in a optical 'double-ridge' waveguide", Electronics Letters, Vol. 25, pp. 746-748, 1989.

4.26 E.C.M. Pennings, "Small-radii curved ridge waveguides for integrated optics", Procs. Eur. Conf. on Opt. Integr. Systems at Amsterdam, The Netherlands, September 25-28, 1989, paper W4.

4.27 D.K. Doeksen, F.H. Groen, Y.S. Oei, J.W. Pedersen, E.C.M. Pennings, H. van Brug and J.J.G.M. van der Tol, "Low propagation and bending losses in LPE-grown InGaAsP ridge waveguides", Procs. Eur. Conf. on Opt. Integr. Systems at Amsterdam, The Netherlands, September 25-28, 1989, paper T2.

4.28 H. van Brug, F.H. Groen, Y.S. Oei, J.W. Pedersen, E.C.M. Pennings, D.K. Doeksen, and J.J.G.M. van der Tol, "Low-loss straight and curved ridge waveguides in LPE-grown GaInAsP", Electronics Letters, Vol. 25, pp. 1330-1331, 1989.

4.29 B.H. Verbeek, E.C.M. Pennings, J.W.M. van Uffelen, and P.J.A. Thijs, "Fabrication and analysis of low-loss InGaAsP/InP optical waveguides with extremely small bends", Procs. 15th Europ. Conf. on Opt. Communication ECOC'89, Sept. 10-14, 1989, Gothenburg, Sweden, paper PDB-9.

4.30 H. Blok, J.M. van Splunter, and H.G. Jansen, "Leaky-wave modes and their role in the numerical evaluation of the field excited by a line source in a non-symmetric, inhomogeneously layered, slab waveguide", Applied Scientific Research, Vol. 41, pp. 233-236, 1984.

4.31 Lord Rayleigh, "The problem of the whispering gallery", the London, Edinburgh, and Dublin philosophical magazine and journal of science, Series 6, Vol. 20, pp. 1001-1004, 1910.

4.32 E.-G. Neumann, "Curved dielectric optical waveguides with reduced transition losses", IEE Proceedings, Vol. 129, Pt. H, pp. 278-280, 1982.

4.33 S. Sheem and J.R. Whinnery, "Guiding by single curved boundaries in integrated optics", Wave Electronics, Vol. 1, pp. 61-68, 1974/75.

Chapter 5

Directional couplers

This chapter describes the design and analysis of short directional couplers with reduced sensitivity to fabrication tolerances. Modifications to the conventional Two-Mode Interference (TMI) coupler, aimed at improving coupler performance and reproducibility, are discussed and experimentally tested. A new type of coupler, which we call Multi-Mode Interference (MMI) coupler, is introduced. This coupler combines good performance with excellent reproducibility. Finally, we describe a first experiment with a phase-diversity coupling network for reducing receiver sensitivity to laser phase noise.

5.1 Introduction

Directional couplers are important elements in many optical circuits. The most frequently employed coupler-type (figure 5.1a) consists of two parallel waveguides with a small gap between them. Tolerance requirements for the fabrication process are extremely strict because the coupling strength depends exponentially on the gap width. For components requiring good control of the coupling between different ports, the only practical way to meet these

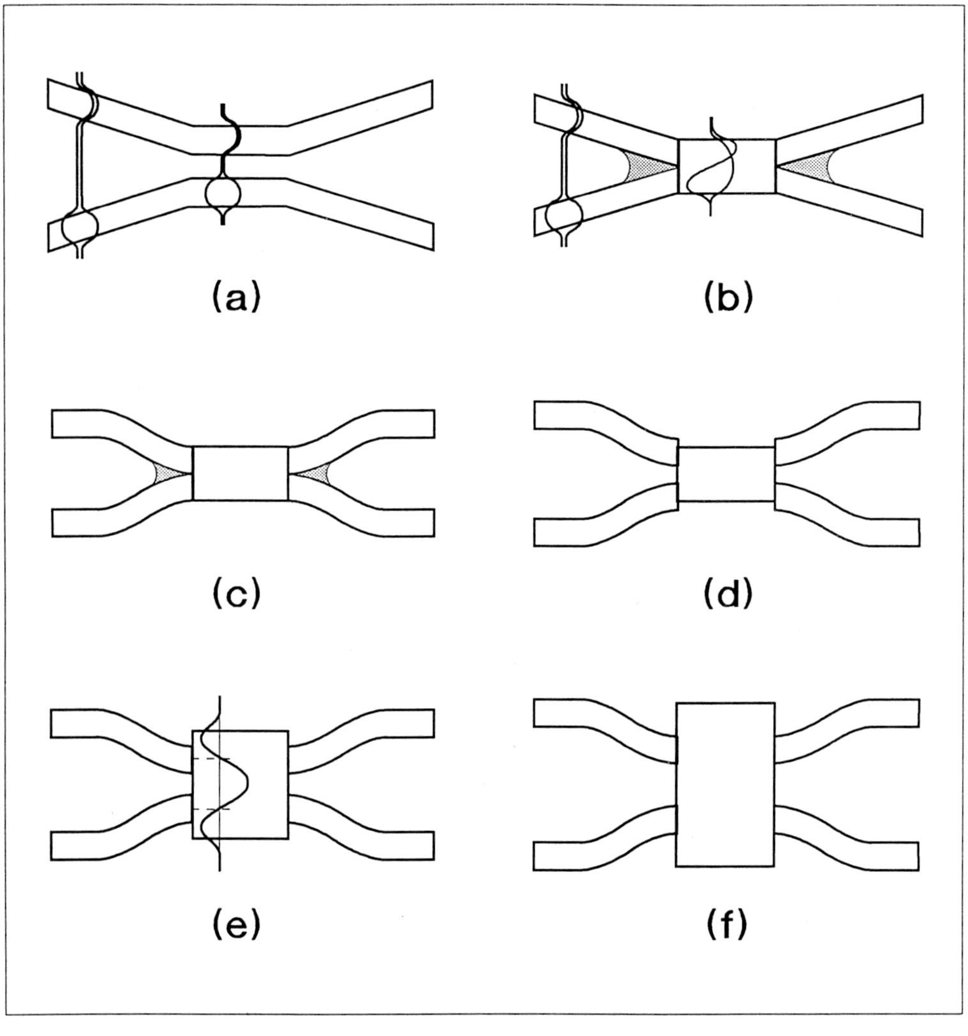

Figure 5.1 *Modifications to the classical coupler scheme for improving coupling reproducibility.*

a) The classical (weakly coupled) directional coupler,
b) TMI-coupler with straight Y-junctions,
c) TMI-coupler with curved Y-junctions,
d) TMI-coupler with increased gap,
e) TMI-coupler with MMI-section,
f) MMI-coupler.

requirements is by including a tuning or trimming facility. This complicates component design and restricts the choice of materials.

The Two-Mode Interference (TMI) coupler (also known as BOA: Bifurcation Optique Active [1]) is a more process-tolerant alternative to the coupler type discussed above. It can be considered as a conventional directional coupler in which the gap width is reduced to zero, as shown in figures 5.1b and 5.1c, thereby creating a coupler section with twice the width of the access waveguides. It is called the TMI-section because it usually supports two modes. TMI couplers have been employed in electrically controlled switches [1,2], polarization splitters [3] and wavelength demultiplexers [4-7].

TMI-couplers have three advantages for use as power splitters. The length of the couplers is small, they are less sensitive to fluctuations of the fabrication process and they exhibit less polarization dependence. Experiments carried out in our laboratory by Schalkwijk [9], showed a considerable reduction of the spread in coupler performance. Residual spread was, however, still too large for the coupler to be applied without a tuning or trimming facility.

Most of the irreproducibility appeared to be caused by the Y-junctions at both sides of the coupler. Due to the finite resolution of the lithographic process the gap between the two Y-junction branches is partially filled in a poorly controllable manner (the shaded regions in figures 5.1b and 5.1c). The poorly reproducible contribution of these junctions to the total response degrades coupler performance.

Several modifications are possible in order to improve the reproducibility. Application of curved access waveguides (figure 5.1c), while not affecting the incoming (or outgoing) angles, reduces the length of the coupling zone in the Y-junctions so that their contribution to the total response is reduced as is, consequently, the effect of fluctuations. If the bending radius of the access waveguides is increased, the mode profile will shift towards the outer edge of the bend. In order to maintain a good field match at the junction with the TMI-section, the access waveguides should be shifted towards the outer edges of the TMI-section, thereby creating a gap, as shown in figure 5.1d. If this gap is greater than the resolution of the lithographic process it will no longer be filled, thus eliminating one source of irreproducibility.

The maximal width of the gap in the modification described above is restricted to twice the shift of the mode profile in the curved waveguide. The gap width can be further increased if a wider (multimode) coupler section is employed, as depicted in figure 5.1e. If the access waveguides are positioned such that their mode profiles are centered around the zeroes of the second-order mode in the wide coupler section, this mode will not be excited and the coupler will operate in a two-mode interference regime (if higher order modes are not excited). Veerman [9] found that the coupling contribution from Y-junctions in these modified couplers is considerably reduced while the coupler reproducibility enhanced. Experimental results revealed that coupler performance improves with increasing coupler-section width, despite the occurrence of higher-order modes. Pennings [10] found that this phenomenon is due to the self-imaging properties of multi-mode waveguides, a phenomenon which appeared to have been suggested already in 1973 by Bryngdahl [11] and experimentally demonstrated and patented by Ulrich [12,13]. Employment of couplers based on this principle, which we called Multimode-Interference (MMI) couplers (figure 5.1f), is most promising with regard to the development of reproducible power splitters.

In this chapter experiments will be described incorporating the modifications to the TMI-coupling principle mentioned above, and their application to a phase-diversity coupling network for reducing receiver sensitivity to laser phase noise.

5.2 Two-Mode Interference Couplers

5.2.1 Power division analysis

In synchronous couplers the guided power is periodically transferred between the coupled waveguides, as illustrated in figure 5.2. If the upper waveguide of the coupler is excited with unit power at $z=0$, the power in the upper (u) and lower (l) waveguide can be described as

$$P_u(z) = cos^2(\kappa z), \quad (1a)$$

$$P_l(z) = sin^2(\kappa z), \quad (1b)$$

respectively, in which κ is called the coupling coefficient. The length L_c:

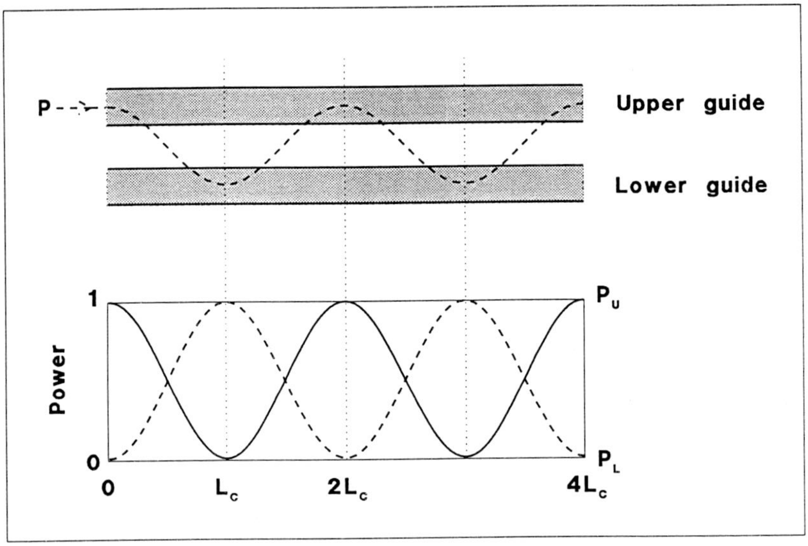

Figure 5.2 *Periodic power transfer in a coupled waveguide system.*

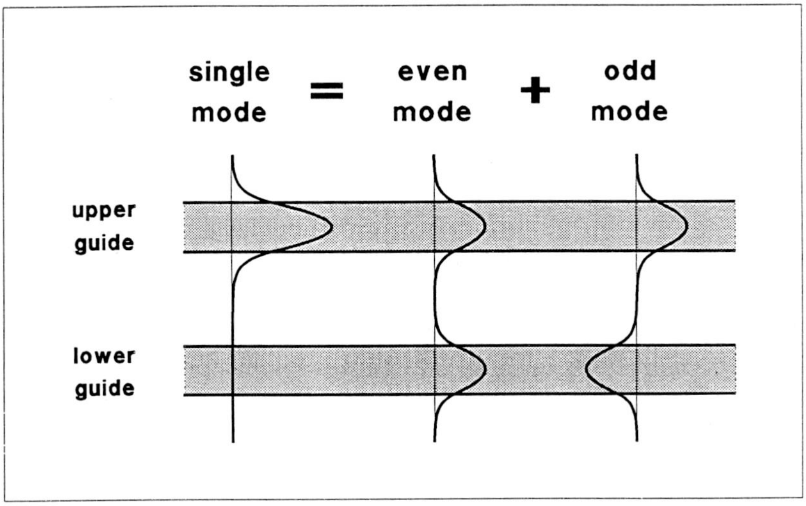

Figure 5.3 *Decomposition of the mode of a single waveguide into the even and odd mode of the two-waveguide system.*

$$L_c = \frac{\pi}{2\kappa}, \tag{2}$$

at which all power is transferred from one waveguide to the other is called the coupling length.

For weakly coupled waveguides the coupling coefficient κ can be computed using Coupled-Mode Theory [14]. A method which is relevant to both weakly and strongly coupled waveguide systems is the System-Mode concept. For weakly coupled waveguides the fundamental system mode approaches the sum-field of the unperturbed modes of the individual waveguides, it is called the even system mode. The first-order system mode approaches the difference of the individual modes (i.e. the individual modes having opposite signs), it is called the odd system mode. Figure 5.3 illustrates how a signal applied to one waveguide can be decomposed into the even and odd system mode. Excitation of a single port thus results in excitation of both system modes with equal amplitude and equal or opposite sign.

If the coupler geometry changes sufficiently slowly in the propagation direction, the system-modes will gradually adapt their profiles to the changing geometry without losing power to other modes. The mode profile is said to change adiabatically[1].

---

1) An adiabatically changing waveguide structure is a structure (e.g. a taper) which changes so gradually that the mode(s) under consideration can adapt their mode profiles to the changing structure without loosing power to other modes (no mode conversion). Consequently, in adiabatically changing structures all power guided in a mode will remain guided in that mode along the whole structure.

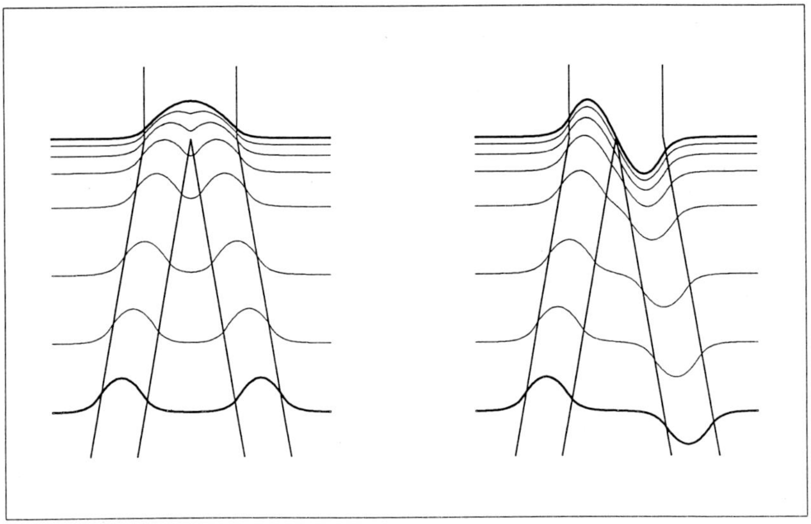

Figure 5.4 *Evolution of the modes of a weakly coupled system on reducing the gap width to zero.*

Figure 5.4 shows how the modes of a weakly coupled system degenerate into the modes of a single waveguide width double width: the TMI-coupler. At the output of the TMI coupler the same process occurs in the opposite direction; the modes in the wide waveguide section will transform into the even and odd mode of the uncoupled system. Power coupled to a system mode at the input of the coupler will remain unchanged while propagating through the coupler and the coupling phenomenon can be described as the interference between the two system modes. From this consideration it is clear that the classical coupler is based on two-mode interference as well.

If the two system modes have the same sign they will interfere constructively in one waveguide and destructively in the other. If the modes have opposite signs the interference will occur just the other way around. Total power transfer thus corresponds to a phase difference of 180° between the two system modes, and an equal division of power to a 90° phase difference. The length L for which $(\beta_0-\beta_1)L$ equals π is thus seen to equal the coupling length, as defined in equation 2, from which the following relation between κ and the difference $\Delta\beta_{01}$ in propagation constants is found:

$$\kappa = \tfrac{1}{2}(\beta_0 - \beta_1) = \tfrac{1}{2}\Delta\beta_{01}. \tag{3}$$

Using this relation the coupling coefficient can be computed for a variety of coupled waveguide systems. An important advantage of the System-Mode approach is that it applies both to strongly and weakly coupled waveguides, including the degenerate case of a zero inter-waveguide gap, which can not be analyzed with Coupled-Mode Theory.

From formula 5.2 it may be seen that increased coupling leads to a reduced coupling length, so that a TMI coupler will be considerably shorter than a weakly coupled system realized in the same waveguide structure.

The analysis of a TMI-coupler, including Y-junctions, employing the System-Mode approach means, essentially, integrating the phase difference which the system modes experience while propagating through the coupler. If the upper waveguide (port 1) is excited with unit power, a simple calculation reveals that the power at the upper and lower output port (ports 3 and 4, respectively), depends on the accumulated phase difference Φ according to:

$$P_3 = \cos^2(\tfrac{1}{2}\Phi), \tag{4a}$$

$$P_4 = \sin^2(\tfrac{1}{2}\Phi), \tag{4b}$$

with

$$\Phi = \int \Delta\beta_{01}(z)\, dz = \int \{\beta_0(z) - \beta_1(z)\}\, dz, \tag{5}$$

in which the integration is along the whole coupling zone. For adiabatically changing waveguide structures the local values of β_0 and β_1 can be computed with a transfer-matrix method by considering the structure as piecewise cylindrical. Whether a geometry may be considered to vary adiabatically can be globally tested using a beam-propagation method.

5.2.2 Insertion loss analysis

In addition to the normal propagation loss, two mechanisms contribute to coupler loss. The first one is the directional change which the propagating light experiences on crossing the junction between the access waveguide and the TMI-section. The second one is the discontinuity in the Y-junction geometry resulting from of partial filling of the inter-waveguide gap. Both effects will be shortly discussed.

A. Insertion loss at a directional change

If the Y-junction angle is sufficiently small, the modes in the two branches will adapt themselves to the increasing proximity of the other waveguide and propagate without loss into the output waveguide. A worst-case approximation of the loss occurring at the junctions between the access waveguides and the TMI-section is therefore obtained by ignoring this adaption and considering the loss to be caused by a discontinuous change of the propagation direction. In this approach the coupling loss is found from the overlap between the modal field in the access waveguide and the sum-field of the excited modes in the TMI-section. The latter field is assumed to be identical to the former field except for a factor $exp(-jk_y y)$, in which $k_y = \beta \sin\theta_Y \simeq \beta\theta_Y$, and θ_Y is the half angle between the Y-junction branches at the end of the junction (see figure 5.5). The exponent accounts for the phase variation along the coupling plane due to the skew incidence of the exciting mode.

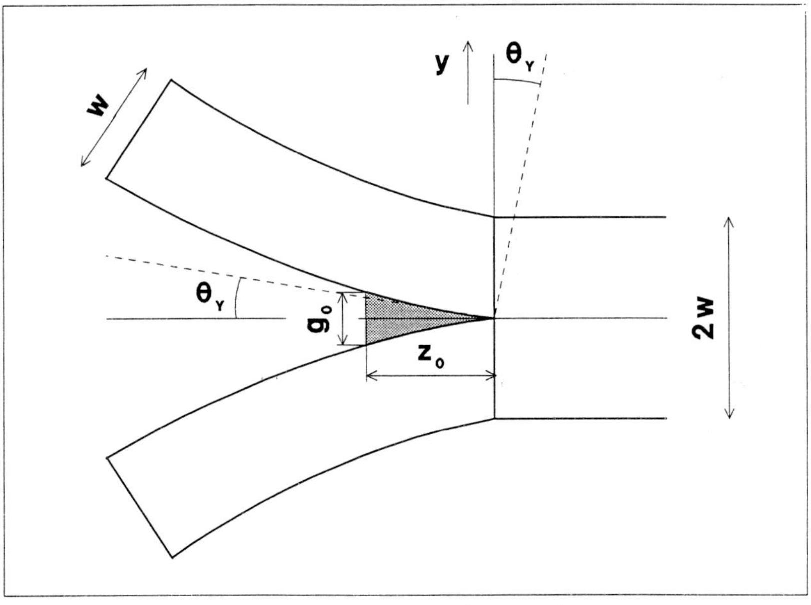

Figure 5.5 Geometry of a Y-junction with a partially filled gap.

The resulting coupling integral is a Fourier integral. If the modal field of the waveguide is approximated by its equivalent Gaussian distribution (see chapter 3 section 2.1B), the coupling efficiency η_j of a single junction will have a Gaussian angle dependence:

$$\eta_j(\theta_Y) = e^{-\theta_Y^2/(2\theta_o^2)}, \tag{6a}$$

with

$$\theta_o = \frac{\lambda}{\pi w_o} = \frac{\sqrt{2\pi}}{\beta w_e}. \tag{6b}$$

For small values of θ_Y the maximum coupling loss (of two junctions) may be estimated from:

$$\eta_{TMI}(\theta_Y) = \eta_j^2(\theta) \simeq 1 - (\beta w_e \theta_Y)^2/2\pi, \tag{7a}$$

or

$$L_{TMI}(\theta) = 10 \log_{10} \eta_{TMI} \simeq 0.7\, (\beta w_e \theta_Y)^2, \qquad (dB). \tag{7b}$$

For single-mode waveguides with a V-parameter between 2.5 and 3 the effective mode width w_e equals the waveguide width w within ±10%. For a 2-µm wide waveguide with $V=3$, and a wavelength of 0.4 µm within the waveguide, the predicted maximal loss for a junction half angle $\theta_Y = 1°$ amounts to 0.2 dB.

B. Insertion loss at a discontinuity

The partial filling of the gap between the access waveguides causes a discontinuity, as can be seen from figure 5.5. At this discontinuity power will be lost. The coupling loss of the even and odd system modes can be computed by overlapping the corresponding mode profiles on both sides of the discontinuity. There is no mode conversion from even to odd modes and vice versa because of the symmetry of the configuration. Figure 5.6 shows the coupling loss at the discontinuity for both system modes as a function of the filling ratio g_o/w, for a waveguide with $V=3$.

From the figure it can be seen that coupling loss of the odd system mode is negligibly small, which was to be expected because this mode has a low field

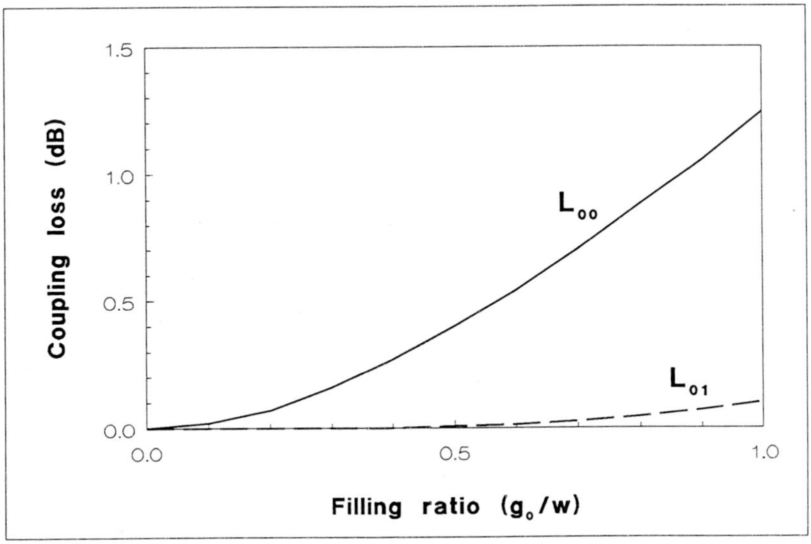

Figure 5.6 Coupling loss of the even (L_{00}) and the odd (L_{01}) system mode at the discontinuity in a partially filled Y-junction.

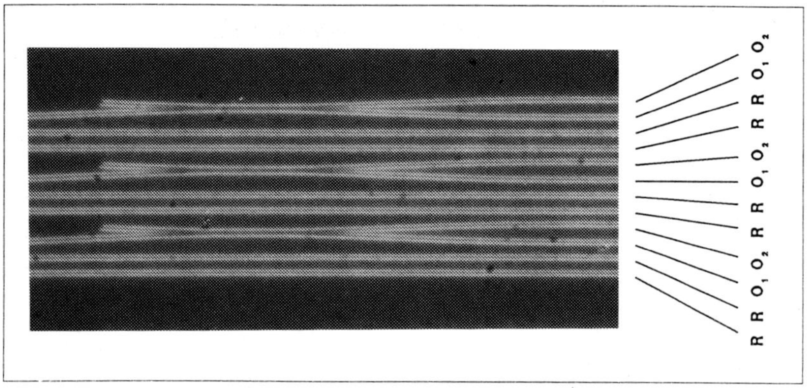

Figure 5.7 *Microscope photograph showing some of the experimental TMI-couplers. Straight channels are included between the couplers in order to serve as references.*

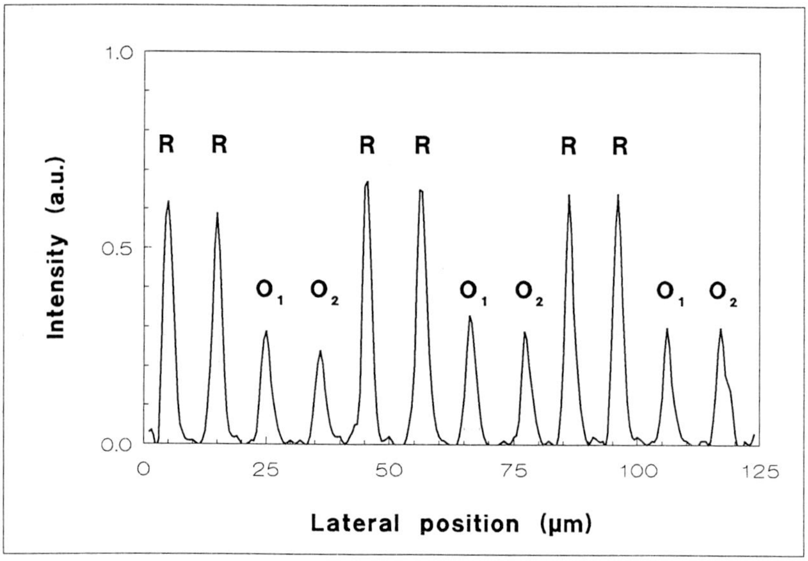

Figure 5.8 *Intensity scan across the output channels at a cleaved end face. The positions of the channels are indicated in figure 5.7.*

intensity at the location of the discontinuity. Therefore, the coupling loss is principally determined by the even mode. If one port is excited, 50% of the power is carried by this mode, and the total coupling loss (for two junctions) due to this phenomenon may be expected to equal the single-junction loss for the even system mode. For a 2 μm wide waveguide with a 1 μm filled gap ($g_o/w=½$) we can thus expect an insertion loss of 0.4 dB. Because of the symmetry of the configuration odd modes will not be excited and the lost power will couple mainly to the second-order mode.

5.2.3 Experimental results

A series of TMI-couplers with varying length was realized with straight and curved waveguides, and Y-junction half angles θ_Y of 0.5° and 1°. Couplers were designed for operation at 633 nm wavelength, with a waveguide width of 2 μm and a lateral V-parameter $V \simeq \pi$. For the curved waveguides a bending radius of 2000 μm was chosen. With this radius the mode shift is 0.1 μm, and the penalty for omitting the offset between straight and curved waveguides is 1%. The difference $\Delta\beta_{01}$ between the fundamental and the first-order mode amounts to 0.0329 μm^{-1} for TE and 0.0341 μm^{-1} for TM-polarization, so that the polarization dependence of the accumulated phase difference Φ will be in the order of 3%.

Figure 5.7 shows a microscope photograph of part of the experimental chip. The couplers were excited with a broad He-Ne laser beam through an input coupling prism, as discussed in chapter 8. Straight waveguides between the

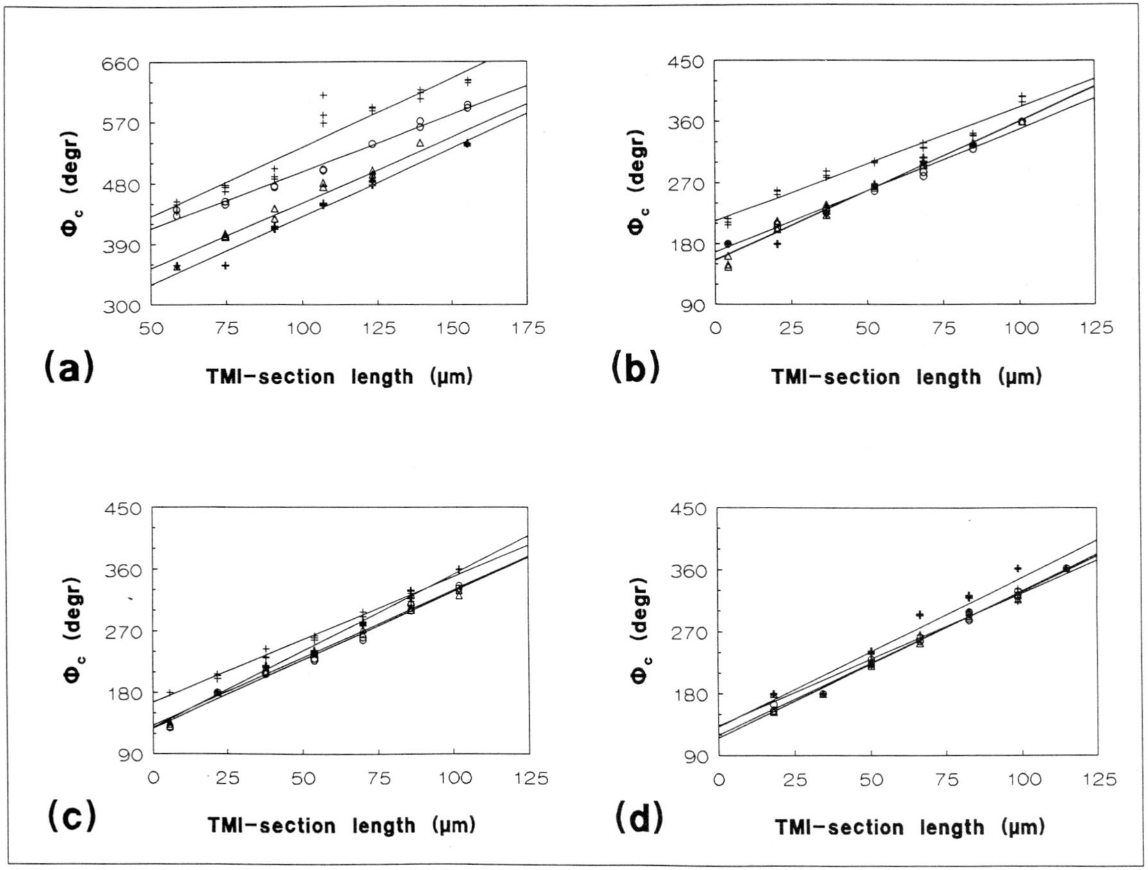

Figure 5.9 *The experimentally determined phase parameter* Φ *as a function of the TMI-section length, for four different coupler geometries:*

a) $\theta_y = \frac{1}{2}°$, $R = \infty$; *b)* $\theta_y = \frac{1}{2}°$, $R = 2000\ \mu m$; *c)* $\theta_y = 1°$, $R = \infty$; *d)* $\theta_y = 1°$, $R = 2000\ \mu m$.

couplers (the reference channels R), which are excited simultaneously with the couplers, were used to indicate the 100% transmission level. In order to excite only one input channel, the second channel of each coupler was not extended into the excitation region, as shown on the photograph. The intensities in the output channels (O_1 and O_2) were measured with a CCD camera on a cleaved end face. Figure 5.8 shows a horizontal intensity scan across the output channels from which the relative channel intensities can be determined by comparing them to the intensities in the reference channels.

The experimental devices were characterized by their power division performance and their insertion loss. Power division was characterized, according to equation 1, by the accumulated phase difference Φ as determined from the ratio between the measured intensities in the output channels with one of the two inputs excited:

$$\Phi = arctan\sqrt{\frac{I_3}{I_4}}. \qquad (8)$$

Insertion loss was determined as

$$L = -10\ log_{10}\ (\frac{I_3+I_4}{I_{ref}}), \qquad (9)$$

with I_{ref} denoting the (averaged) intensity measured at the reference channels.

Figure 5.9 shows the experimentally determined phase parameter Φ for four coupler types with straight and curved *(R=2000 μm)* access waveguides, and 0.5° and 1° Y-junction half-angles. Each graph comprises measurement data collected on four identical series, divided over two substrates. Regression curves are shown for each series.

From these graphs the coupling contribution of the Y-junctions can be inferred as the Φ-value at which the regression curves intersect the vertical axis (i.e. the value of Φ for zero TMI-section length). As expected the Y-junction contributions are maximal for straight access waveguides and a 0.5° half angle: 140°-190° per junction. Curved waveguides reduce this contribution to 80°-105°, a reduction of almost 50%. Increasing the 0.5° angle to 1° leads to an even greater reduction: 60% with straight waveguides and 65% with curved waveguides.

The spread in the data is approximately proportional to the magnitude of the Y-junction contributions, which confirms our supposition that the irreproducibility is caused by these junctions. Figure 5.10 shows the phase contribution of the junctions defined according to formula 5.8. The total phase contribution Φ_Y of the couplers consists of a contribution Φ_T of the taper section which arises from the filling of the gap between the junction branches over a length z_o (see figure 5.5), and a contribution Φ_B of the two-branch section. Predicted results were obtained by computing the integral of equation 5.5 for the structure depicted in figure 5.5. The actual length z_o was measured

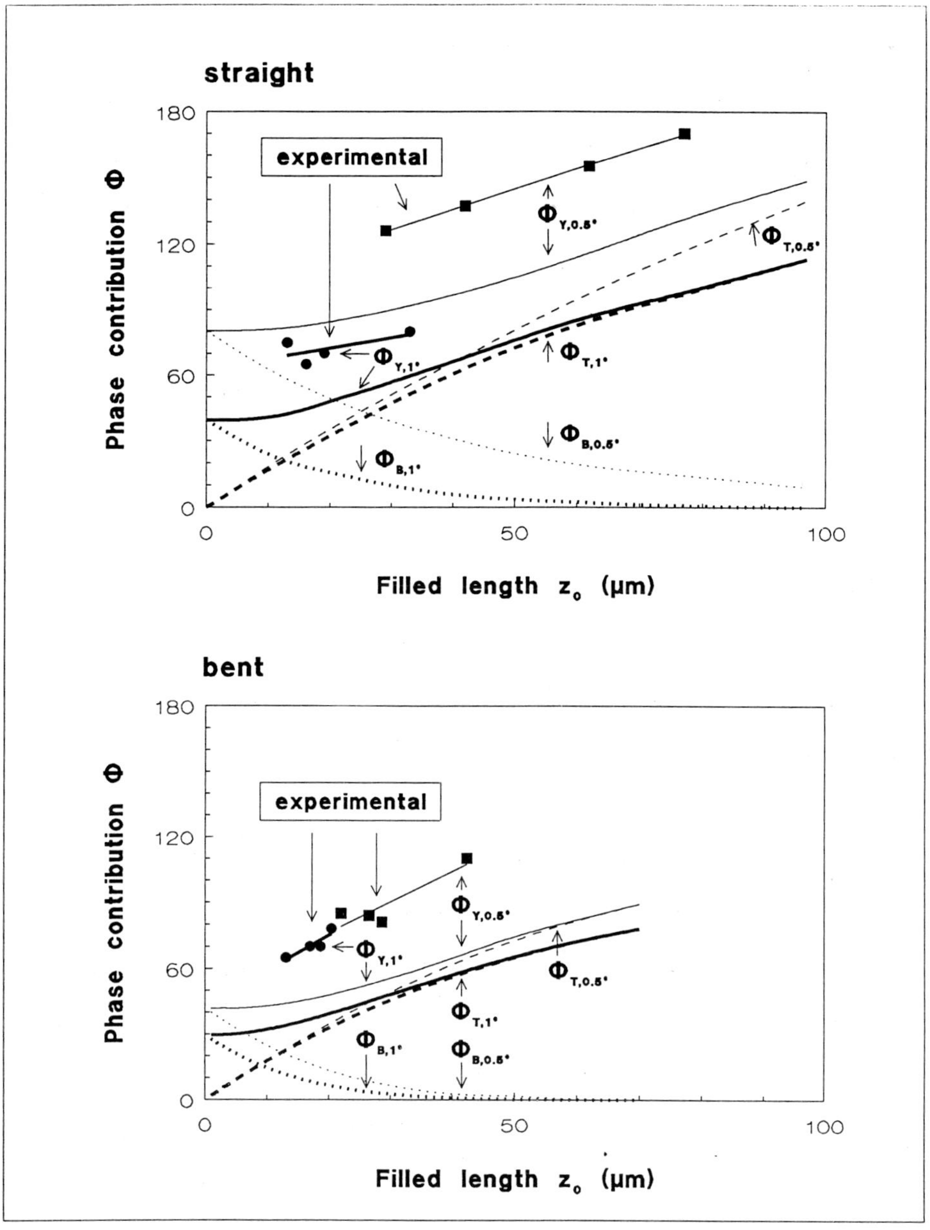

Figure 5.10 The phase contribution Φ_Y of the Y-junctions (predicted and measured), as composed of the contribution Φ_T of the taper section and the contribution Φ_B of the two-branch section. Results are shown for straight (upper figure) and curved Y-branches (lower figure), and Y-junction half angles of 0.5° and 1.0°, respectively, as a function of the experimentally determined length z_o (see figure 5.5).

on the chip with a microscope. Comparison of the measured results with the predicted values reveals that most of the spread in the data is attributable to differences in the length of the filled region. After correction for this effect the difference between theory and experiments is reduced to 30° ± 10°.

Figure 5.11 shows the frequency distribution of the insertion loss, as determined experimentally for the different coupler types. No significant dependence on coupler length was found. Insertion loss varies between 0.6 and 1.2 dB, and is approximately ½ dB higher than expected from the junction filling effect (figure 5.6). Curvature increases insertion loss slightly, probably due to increased mode mismatch. Contrary to our expectation a

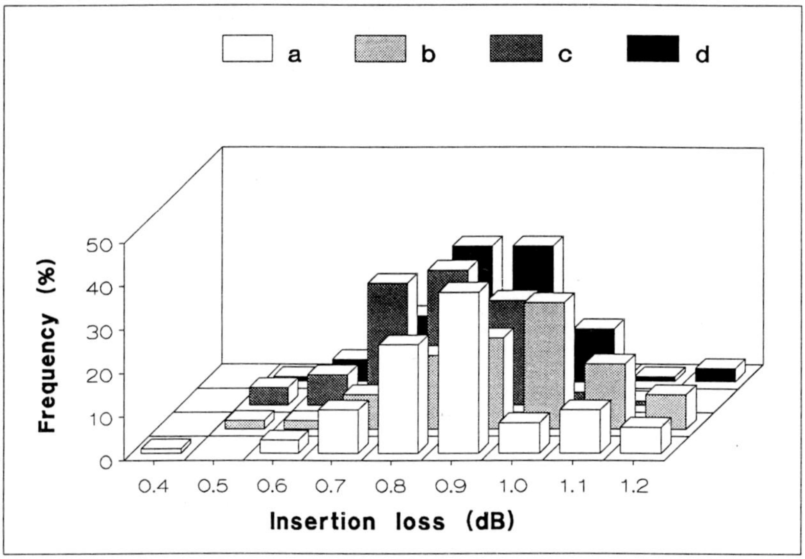

Figure 5.11 Experimentally determined occurrence frequency distributions for the insertion loss of four different coupler types with:

a) $\theta_y = ½°$, $R=\infty$ *b)* $\theta_y = ½°$, $R=2000\ \mu m$
c) $\theta_y = 1°$, $R=\infty$ *d)* $\theta_y = 1°$, $R=2000\ \mu m$

higher value of θ_Y leads to a smaller loss, although the difference is not great ($\simeq$0.1 dB).

It is concluded that TMI-couplers can be realized with insertion losses between 0.5 and 1 dB, and an accumulated phase difference (Φ) reproducibility within ± 10°, assuming that the partial filling of the inter-waveguide gap in the Y-junctions can be eliminated or controlled.

5.3 Modified TMI-couplers

As a first modification we applied the shift of the mode profile occurring in strongly curved waveguides, to move the access waveguides into the opposite direction while retaining the field match at the junction with the wide TMI-section. If the gap between the access waveguides, which is created by this modification, is larger than the resolution of the lithography, it will no longer be filled, thus eliminating a source of irreproducibility. A second modification is to enlarge the width of the central coupler section. Simulation results indicate that this should improve coupler performance. The modifications are depicted in figures 5.1d and 5.1e. Couplers as depicted in these figures can no longer be considered adiabatic: a different analysis is needed to predict coupler response. To analyze these coupler structures we will first neglect coupling effects in the Y-junctions, thereby reducing the problem to the computation of transmission through a track of cascaded waveguides with different geometries, as depicted in figure 5.12. Coupling effects will be considered separately.

5.3.1 Power transfer properties

If mutual coupling effects between adjacent input or output waveguides is neglected, transmission from a single input port to the two output ports of the coupler can be analyzed using the two waveguides tracks depicted in figure 5.12. If we confine ourselves to single mode input and output waveguides,

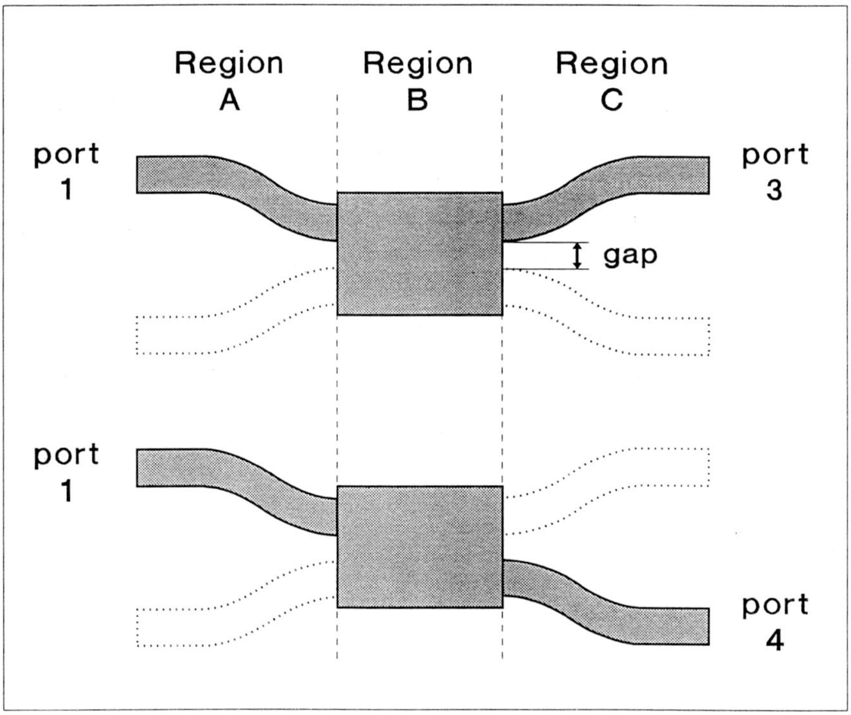

Figure 5.12 Waveguide tracks as used for the analysis of transmission from an input port to the in-line output port or the opposite output port.

the power transfer coefficient t_{ji} of input waveguide i to output waveguide j (both of which are assumed to be monomode) follows as

$$t_{ji} = |C_{CB} P\, C_{BA}|^2 \tag{10}$$

in which C_{CB} and C_{BA} are the overlap matrices of the junctions between regions A and B, and regions B and C, respectively, with elements:

$$c_{pq} = \frac{\int U_{pk} U_{ql}\, e^{j k_y y}\, dy}{\int | U_{pk} |^2\, dy}. \tag{11}$$

Note that the matrices are different for the two configuration depicted in figure 5.12. The functions U_{pk} and U_{ql} denote the profiles of the k-th and the l-th mode in regions p and q, respectively. The exponent $jk_y y = j\beta_{pk}\, y \sin\theta_y$ (θ_y is the Y-junction half angle at the end of the junction (see figure 5.5) accounts for the phase variation across the junction due to the skew incidence of the exciting mode relative to the excited one. The matrix P is the (diagonal) propagation matrix of the TMI-section (region B) with diagonal elements

$$p_i = e^{-j\beta_i L_{TMI}}, \tag{12}$$

in which β_i is the propagation constant of mode i in the TMI-section.

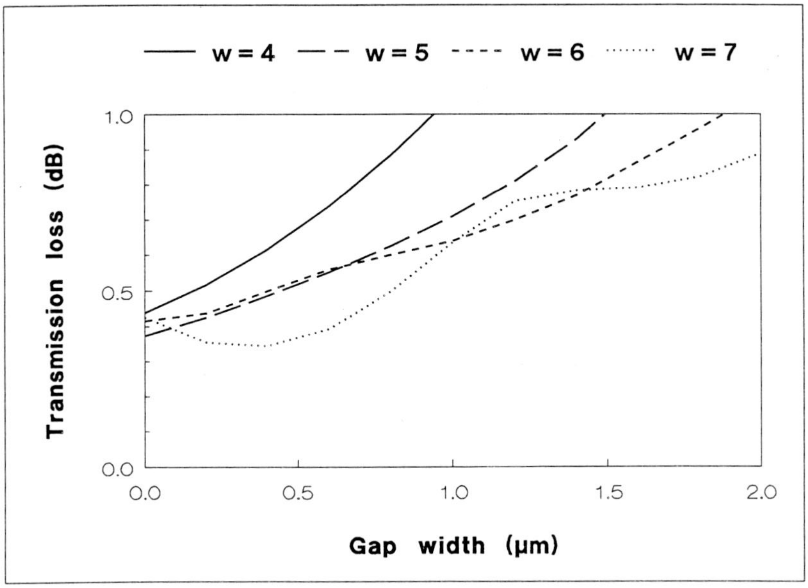

Figure 5.13 *Cross-state transmission loss to the opposite port as a function of the inter-waveguide gap (see figure 5.12) for different values of the coupler section width.*

Figure 5.13 shows the transmission loss for the cross state (i.e. $\Delta\beta_{01}L=\pi$). Results were computed for 2 μm wide access waveguides with a radius of 2000 μm, having a core effective index of 1.5733 at 633 nm wavelength, and an effective index contrast of 0.0086. From the figure it can be seen that transmission loss increases with increasing gap width and decreases with increasing coupler section width. The effect of a wider waveguide separation can thus be compensated for by applying a wider coupler section.

Figures 5.14 and 5.15 show the predicted and the experimentally measured transmission to both output ports as a function of the normalized length L_{Φ} of the coupler:

$$L_{\Phi} = L_{\Phi,B} + 2\Phi_Y \tag{13a}$$

$$L_{\Phi,B} = \Delta\beta_{01} L_B \, , \tag{13b}$$

in which L_B is the length of the center-section (as shown in figure 5.12) and $L_{\Phi,B}$ is its normalized length. The actual value of the phase contribution $2\Phi_Y$ of the Y-junctions can be estimated from the shift which is necessary to fit the predicted curves, as shown in figures 5.14 and 5.15, which do not account for the effect of the Y-junctions, to the experimentally measured results. It can also be determined as the intersection of the regression line through the

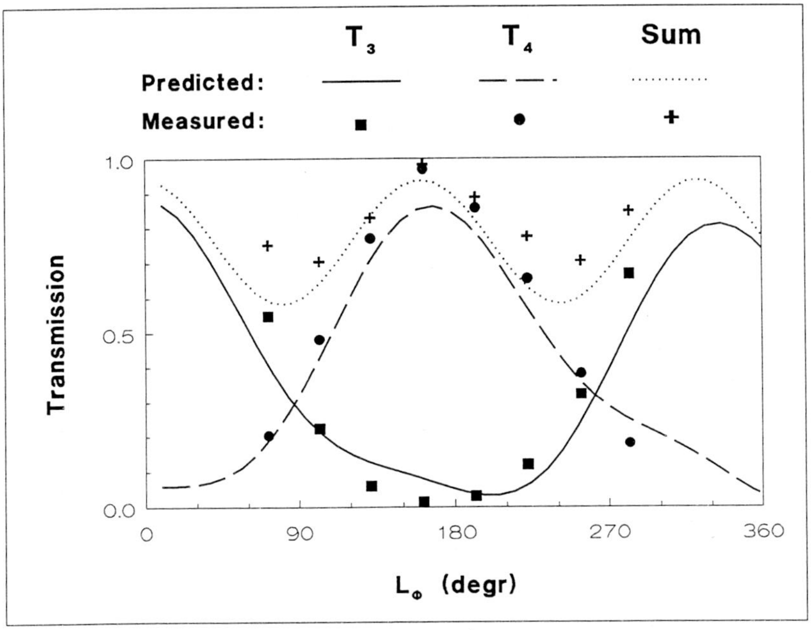

Figure 5.14 *Transmission to both output waveguides (T_3 and T_4) and total transmission (T_3+ T_4) as a function of the normalized coupler length $\Delta\beta_{01}L$. Coupler width is 4 µm, gap width 0.8 µm. Markers indicate experimental results (λ=633 nm).*

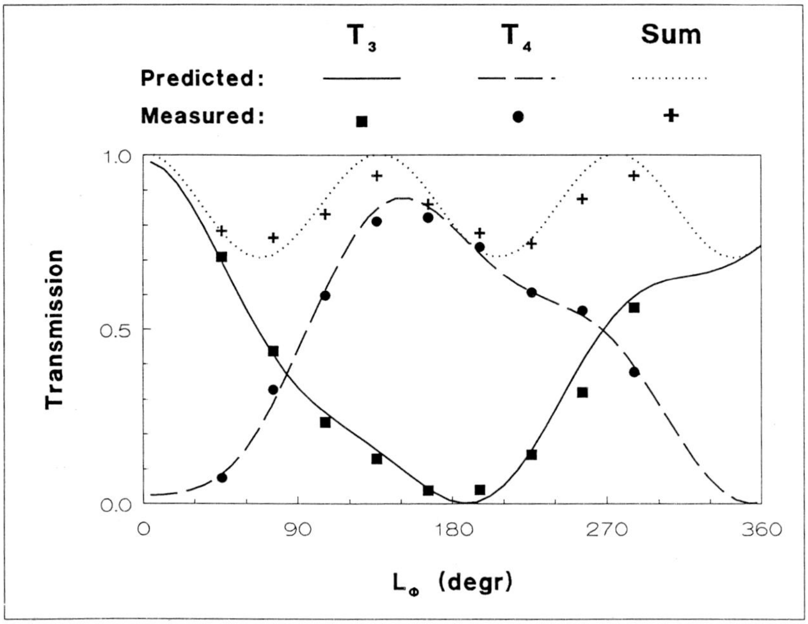

Figure 5.15 The same results as shown in figure 5.14, but for a 7 μm wide coupler with a 1.6 μm wide gap (λ=633 nm).

measured data with the vertical axis ($L_{\Phi,B}=0$), as shown in figure 5.19[1]. A normalized coupler length was introduced in order to facilitate comparison with conventional couplers: 180° and 360° correspond with the cross and the bar state, respectively, 90° and 270° with 3-dB states.

Results are shown for two different coupler types. The first coupler has a 4 μm center-section width and a 0.8 μm inter-waveguide gap. It differs from the TMI-couplers as discussed in the previous section in that an offset is applied

1) It is assumed that the phase contributions of the Y-junctions and the center section are additive. This point is discussed in more detail in the next section (5.3.2).

to the access-waveguides in order to compensate for the modal shift occurring in the bent waveguides. The center section guides three modes as a result of the index contrast requirements for the bends. If the offset is optimized with respect to minimal insertion loss in the 3-dB state, the three guided modes carry 50%, 35%, and 9% of the single-port input power, respectively, so that 6% (per junction) is lost to the radiation field.

The transmission curves deviate slightly from a sinusoidal shape which is due to the occurrence of a second-order mode. The effect of the second-order mode is seen in the slightly non-sinusoidal character of the transmission curves. Insertion loss in the 3-dB states appears to be very high: 2 dB at 270° normalized coupler length and even more at 90° length. The application of a gap to improve reproducibility, is thus paid for with an increased insertion loss, resulting from the reduced field match at both sides of the center section.

Figure 5.15 shows predicted and measured transfer properties for the second coupler type, which has a 7 µm wide coupler section and a 1.6 µm gap. This coupler guides 4 modes of which the first three carry 50%, 42%, and 8% of the single-port input power, respectively. As the sum of these fractions approaches 100%, little power appears to be lost to the radiation field. From the figure it can be seen that very low insertion loss is expected at the second 3-dB point, at 270°. There appears to be a considerable difference between the 90° and the 270° 3-dB states, which is probably due to interference with the second order mode which is responsible for the oscillation of the total power. The effect of this mode on the location of the 3-dB point is small: according to the simulation it occurs at 267.6°, so that it is shifted by approximately 2½°. Application of a wider gap reduces coupling effects, which are not accounted for in the modeling, and is thus expected to yield a better fit between predicted and experimental data. This was indeed observed. The results demonstrate that the modified TMI coupler combines reduced sensitivity to fabrication tolerance with insertion loss as low as 0.5 dB.

To test the present modifications at the longer wavelengths used in modern optical communication systems, we realized a coupler series with 12 μm width and a 1.6 μm gap width for operation at 1550 nm wavelength. Figure 5.16 shows the results. The predicted total power transfer slightly exceeds

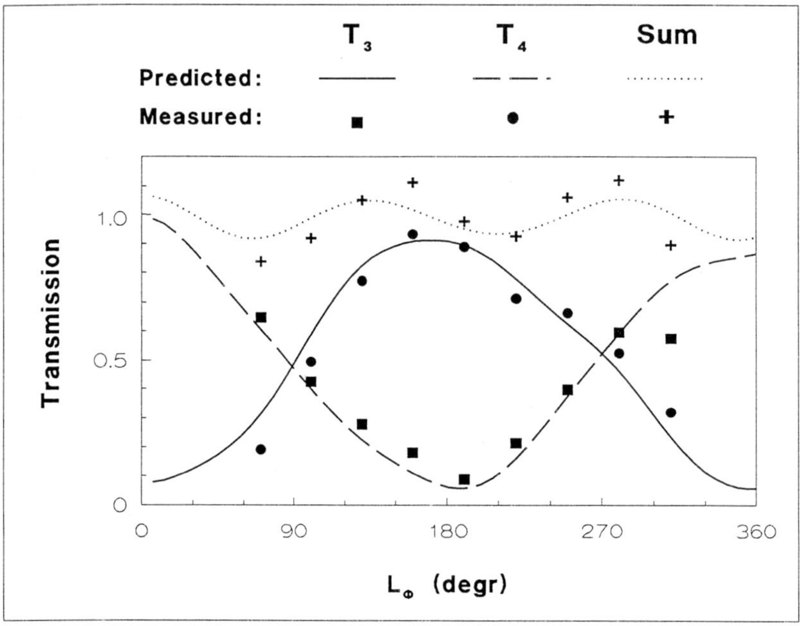

Figure 5.16 The same results as shown in figure 5.14, but for a 12 μm wide coupler with a 1.6 μm wide gap, designed for operation at 1550 nm wavelength.

100% due to neglection of the overlap between the modes of the output ports in the present analysis. Accuracy is reduced correspondingly. The fit between experimental and predicted data is further degraded by the larger magnitude of the coupling effects due to the smaller size of the gap relative to the wavelength and by a greater measurement inaccuracy. Due to the slow response and the strong non-linearity of the infrared vidicon employed, measurement accuracy is not better than ± 1 dB. Taking these accuracy degrading factors into account, it is concluded that long-wavelength coupler performance is comparable to the results obtained at short wavelengths.

5.3.2 Coupling effects in the Y-junctions

Coupling effects in the Y-junctions were analyzed by computing the phase parameter Φ according to equation 5.8 for a coupler with zero center-section length, as shown in figure 5.17. The transmission coefficients T_3 and T_4 were

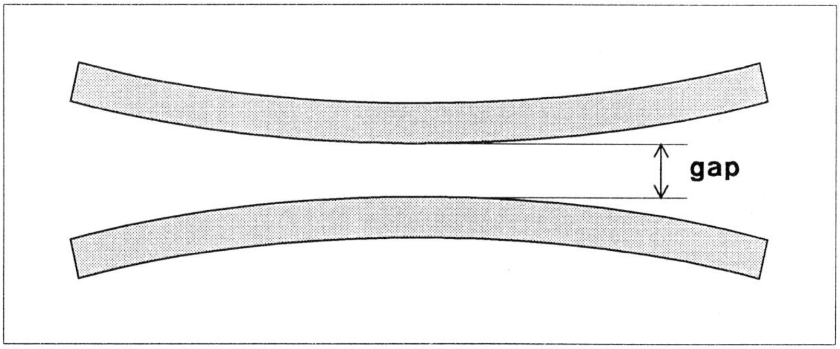

Figure 5.17 Configuration used for computing the phase contribution due to coupling between input and output waveguides; the central section has been omitted.

computed with a beam propagation method[1]. Figure 5.18 shows the computed phase contribution of the applied Y-junctions as a function of the gap-width (computations were performed for 2000 μm bending radius and 633 nm wavelength).

To test the model description we determined the phase parameter Φ, as inferred using formula 5.8 from the experimental results, measured at 633 nm wavelength on a coupler series with a 1.6 μm gap, as described in the next subsection. Figure 5.19 shows the results. From this figure the contribution

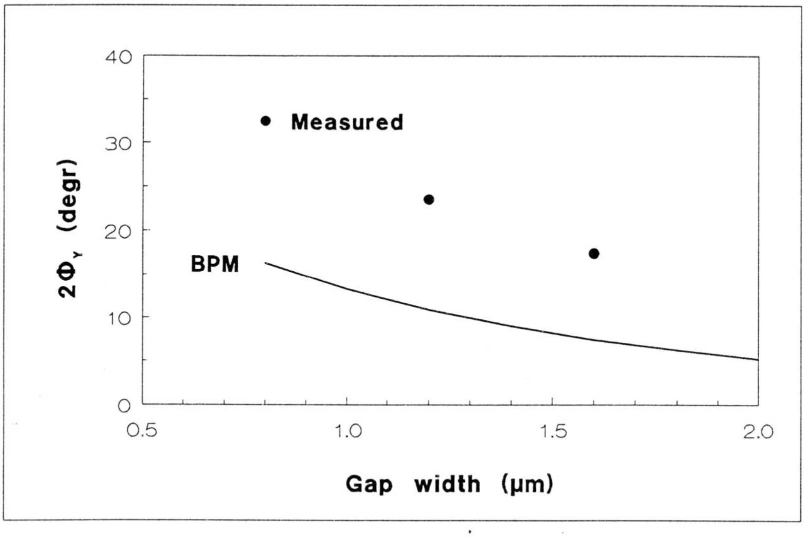

Figure 5.18 *The phase contribution of two Y-junctions with 2000 μm bending radius, as computed with the Beam Propagation Method. Dots indicate experimentally determined results.*

1) The BPM-program developed by G.J.M. Krijnen (Twente University, Enschede, The Netherlands) was used for the computations.

$2\Phi_Y$ of the Y-junctions can be read[1] as the value of Φ for zero central section length (i.e. $L_{\Phi,B}=0$).

The Y-junction contributions, determined in this way, are indicated in figure 5.18. Comparison with the predicted values reveals that, although the shape of the curve follows the predictions, the magnitude of the contribution is

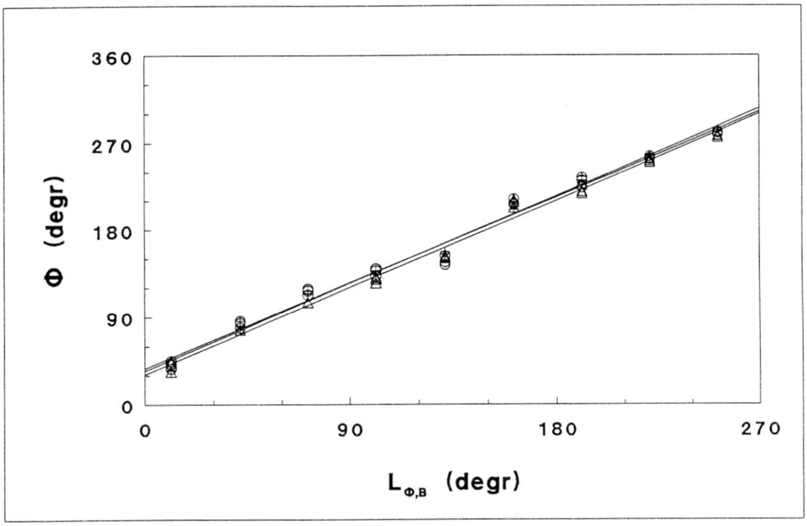

Figure 5.19 The accumulate phase difference Φ for the couplers with 1.6 μm width (see figure 5.15), as determined from the experimental data employing equation 5.8.

1) In multimode coupler sections the dependence of output power on the coupler-section length will become non-sinusoidal, as can be seen from figures 5.14 and 5.15. As a result, the dependence of the phase parameter Φ (equation 5.8) on the coupler length will become non-linear. Further, the phase contributions of the Y-junctions and those of the coupler-sections may become non-additive because of the complex conversion mechanism at the junctions between them. The linear dependence on the coupler length of the data shown in figure 5.19 indicates that for the couplers presently investigated, the non-linear behavior is sufficiently weak to determine the phase contribution of the Y-junctions in the way described.

greater than predicted. A similar difference is observed in figure 5.10. It might be caused by inaccuracy of the Effective Index Method in predicting coupling effects, which are smaller than the propagation constants, from which they are determined, by several orders.

Comparison of figure 5.19 with figure 5.9 shows that modified TMI-couplers exhibit a considerably smaller contribution from the Y-junctions and a better reproducibility: the standard deviation between measured and design values for the phase parameter Φ is within 10°, and part of this deviation will be caused by the non-linearity mentioned in the footnote on the previous page.

It is thus concluded that modified TMI-couplers can be realized with insertion losses in the order of 0.5 dB and a coupling phase parameter reproducibility within ± 10°. Further, it is noted that coupler performance increases with increasing coupler width.

5.4 Multimode-interference couplers

Simulations and experiments on Modified TMI-couplers indicate that coupler performance improves with increasing coupler width. Pennings [10] discovered that this is due to the self-imaging properties of multi-mode waveguides, a phenomenon which had already been suggested in 1973 by Bryngdahl [11] and experimentally demonstrated and patented by Ulrich

[12,13]. A short explanation concerning couplers based on strongly multi-moded waveguides, which we called MMI-couplers, will be given in the following paragraphs. Two versions will be discussed, a longer fully self-imaging version (3L MMI-coupler), and a shorter one, which is also self-imaging if the access waveguides are properly positioned.

5.4.1 3L MMI-couplers

Figure 5.20 illustrates the dependence between the longitudinal and the transverse propagation constants in strongly multimoded waveguides. In such waveguides the lower order modes will be almost completely confined so that their lateral mode profiles will contain an integer number of half periods

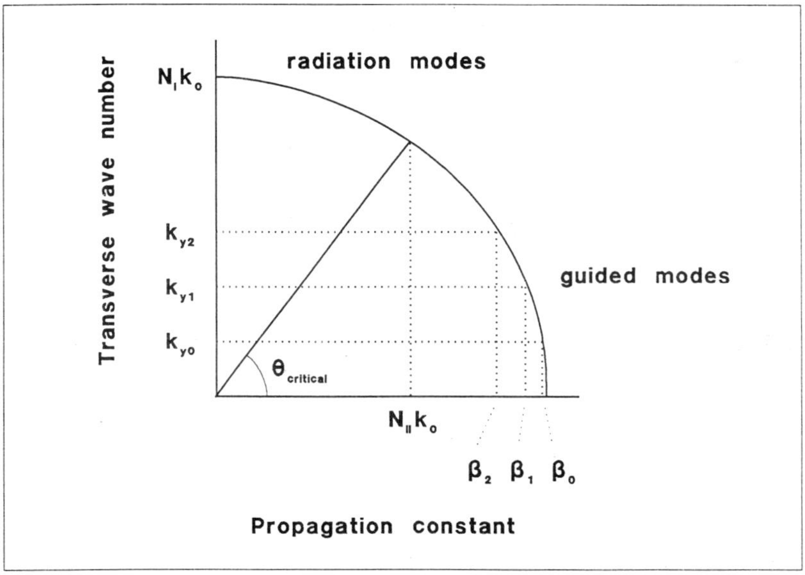

Figure 5.20 The relation between the transverse and the longitudinal propagation constant in a strongly multimoded waveguide.

within the waveguide. Therefore, the higher-order mode profiles are spatial harmonics of the fundamental mode profile and their transverse propagation constants k_{yi} will be integer multiples of the fundamental one (i.e. $k_{yo} \simeq \pi/w$):

$$k_{yi} \simeq (i+1)\, k_{yo}. \tag{14}$$

The corresponding longitudinal propagation constant β_i follows as:

$$\beta_i = \sqrt{N_I^2 k_o^2 - k_{yi}^2} \simeq N_I k_o - \tfrac{1}{2}(i+1)^2 \frac{k_{yo}^2}{N_I k}, \tag{15}$$

which can be rewritten as (by substituting $N_I k_o - \frac{1}{2} \frac{k_{yo}^2}{N_I k} = \beta_0$):

$$\beta_i \simeq \beta_0 - \frac{i\,(i+2)}{3} \Delta\beta_{01}, \tag{16a}$$

with

$$\Delta\beta_{01} = \beta_0 - \beta_1 \simeq \frac{3\pi^2}{2 N_I k_o w^2}. \tag{16b}$$

To analyze the imaging properties of a multimode waveguide section we decompose the exciting field $U_i\,(x)$ at the input of the waveguide into its modal components, propagate them through the waveguide and reconstruct the field at the output of the waveguide as the sum of the modes. The shape of the sum field is determined by the phase difference

$$\Phi_{L,i} = (\beta_i - \beta_0)\, L, \tag{17}$$

of the modes, relative to the fundamental one, after travelling through the waveguide with length L. In our analysis it will prove sufficient to treat the modes as two classes: the even and the odd modes. We can therefore limit ourselves to the decomposition of the exciting field into its even and odd parts:

$$U_i(y) = U_e(y) + U_o(y). \tag{18}$$

To simplify the analysis coupler lengths will be expressed in terms of the coupling length $L_c = \pi/\Delta\beta_{01}$, i.e. the length required for 100% power transfer in a conventional coupler.

We shall first consider a coupler section with a length $3L_c$. For such a length the relative phase transfer Φ_3 of the different modes forms the following series[1] (after substraction of multiples of 2π):

$$\Phi_3 = \{0,\pi,0,\pi,...\}, \tag{19a}$$

so that the odd modes experience 180° phase shift relative to the even ones. For the field U_3 following a length $3L_c$ we thus find:

$$U_3(y) = U_e(y) - U_o(y) = U_e(-y) + U_o(-y) = U_i(-y), \tag{19b}$$

1) The numbers between braces denote the phase transfer for modes 0,1,2,3,...

i.e. the field distribution is inverted with respect to the symmetry axis of the coupler, ***independent of its shape*** (as long as it can be decomposed completely into a set of guided modes). A multimode section with length $3L_c$ thus acts as a generalized cross coupler.

In the same way we find for a coupler length $L=6L_c$:

$$\Phi_6 = \{0,0,0,0,...\}, \tag{20a}$$

and

$$U_6(y) = U_e(y) + U_o(y) = U_i(y). \tag{20b}$$

Such a coupler reproduces the field at its input, and thus acts as a coupler in the bar state.

For a coupler with length $L = 3*(½L_c)$ we find the following results:

$$\Phi_{3/2} = \{0,\frac{3}{2}\pi,0,\frac{3}{2}\pi,...\}, \tag{21a}$$

and

$$U_{3/2}(y) = U_e(y) - jU_o(y) = \frac{1}{\sqrt{2}}\{U_i(y)\,e^{-j\frac{1}{4}\pi} + U_i(-y)\,e^{j\frac{1}{4}\pi}\}. \tag{21b}$$

The field at the output of the waveguide is seen to consist of the original field and its inverse, attenuated by a factor 2 and with a relative phase difference of 90°. The waveguide thus acts as a generalized 3-dB hybrid.

For a length $L = 3*(\frac{3}{2}L_{\pi})$ we find, finally:

$$\Phi_{9/2} = \{0,½\pi,0,½\pi,...\}, \tag{22a}$$

and

$$U_{9/2}(y) = U_e(y) + jU_o(y) = \frac{1}{\surd 2}\{U_i(y)\, e^{j¼\pi} + U_i(-y)\, e^{-j¼\pi}\}. \tag{22b}$$

The only difference from the previous case is that the sign of the phase difference between the two outputs has changed.

The properties of Multi-Mode Interference sections can thus be summarized as follows:

* **An MMI section with length 6L$_c$ reproduces the input field (self-imaging).**

* **An MMI section with length 3L$_c$ inverts the input field.**

* **MMI sections with lengths 3*½L$_c$ and 9*½L$_c$ divide the input power equally between a direct and a ±90° shifted inverse image.**

These properties are periodic in L with period $6L_c$. The three cases are depicted symbolically in figure 5.21. Couplers with section lengths as indicated above will be denoted as 3L MMI-couplers, because their length equals

3 times the length required in the TMI-regime. The advantage of 3L MMI-couplers over TMI-couplers is that the coupling properties are no longer dependent on the shape of the input field. We are thus free in choosing the width, the curvature and the position of the input waveguides, and may choose these positions such that coupling between the access waveguides is completely eliminated.

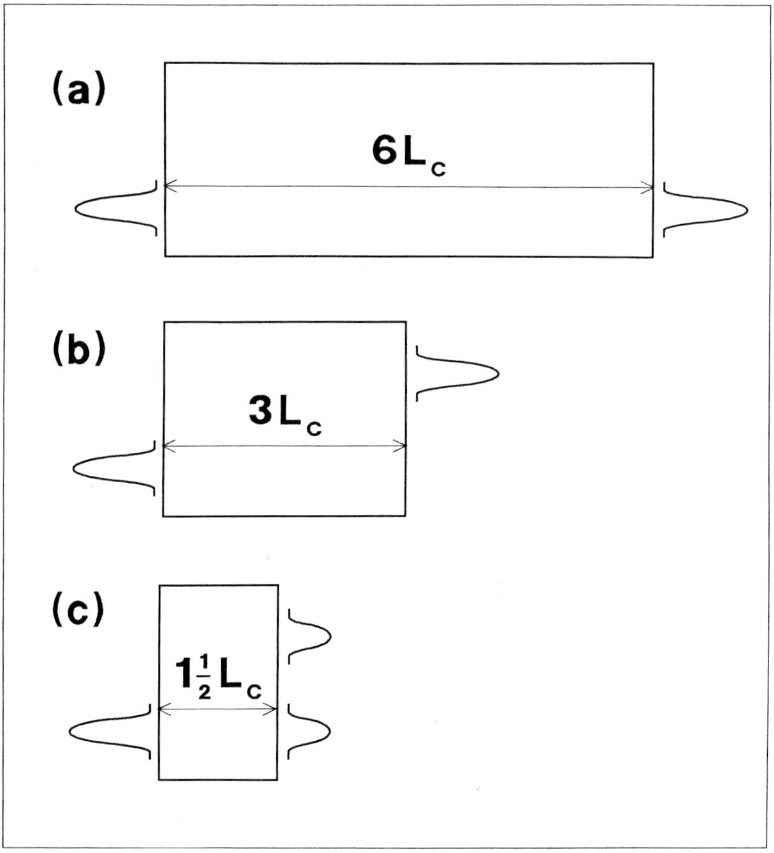

Figure 5.21 Elementary operation of 3L MMI-couplers in (a) the bar state, (b) the cross state, and (c) the 3-dB state.

The resolution of the imaging process is determined by the spatial frequency of the highest order guided mode. This is directly related to the numerical aperture of the waveguide. High resolution requires a high numerical aperture, similar to imaging lenses. A high numerical aperture is obtained by increasing the effective index contrast. It should be noted that for modes close to cut-off the propagation constant will deviate from the idealized relation (equation 5.16) on which the imaging properties are based. Input fields with mode spectra extending close to cut-off will therefore experience distortion, which degrades coupler performance.

The observation that coupler performance improves with increasing coupler width is easily explained in terms of the imaging process: in wider waveguides the number of guided modes increases and the propagation constants of the modes carrying the signal power come closer to the required relation as described in equation 5.16. The observation that 3-dB performance at the 270°-point is better than at the 90°-point (figures 5.14, 15, and 16) is fundamental, and a direct consequence of the imaging properties of 3L couplers.

5.4.2 1L MMI-couplers

In this subsection the properties of MMI-couplers with the same length as conventional couplers will be considered. For couplers with lengths L_c, $2L_c$, $\frac{1}{2}L_c$, and $\frac{3}{2}L_c$ we find, comparable to the longer versions (see equations 5.19-22):

$$\Phi_1 = \{0,\ \pi,\ *,\ \pi,\ 0,\ *,\ 0,\ \pi,\ *,\ \ldots\}, \tag{23a}$$

$$\Phi_2 = \{0,\ 0,\ *,\ 0,\ 0,\ *,\ 0,\ 0,\ *,\ \ldots\}, \tag{23b}$$

$$\Phi_{1/2} = \{0,\ \tfrac{1}{2}\pi,\ *,\ \tfrac{1}{2}\pi,\ 0,\ *,\ 0,\ \tfrac{1}{2}\pi,\ *,\ \ldots\}, \tag{23c}$$

$$\Phi_{3/2} = \{0,\ \tfrac{3}{2}\pi,\ *,\ \tfrac{3}{2}\pi,\ 0,\ *,\ 0,\ \tfrac{3}{2}\pi,\ *,\ \ldots\}, \tag{23d}$$

in which the asterix denotes a fraction different from $\frac{1}{4}\pi$, $\frac{1}{2}\pi$ or $\frac{3}{2}\pi$. The corresponding modes will thus disturb the imaging mechanism. If these modes, with order 2,5,8,... are not excited, the even and odd modes will experience the same phase shifts as in the 3L sections. We thus come to the following conclusion:

*** 1L MMI-Couplers behave in a manner identical to 3L MMI-couplers if the input field is positioned such that the modes of order 2,5,8,... are not excited.**

Figure 5.22 shows the mode-profiles of modes 2,5,8,... in a strongly multimode waveguide. In order not to excite the second-order mode, the input field should be centered around a zero of this mode profile. The profile of the mode being anti-symmetrical around this point, it will not be excited by a symmetrical input field. Because the zeroes of modes 2, 5, and 8 almost coincide, the excitation coefficients of these modes will automatically be small if the second-order mode is not excited.

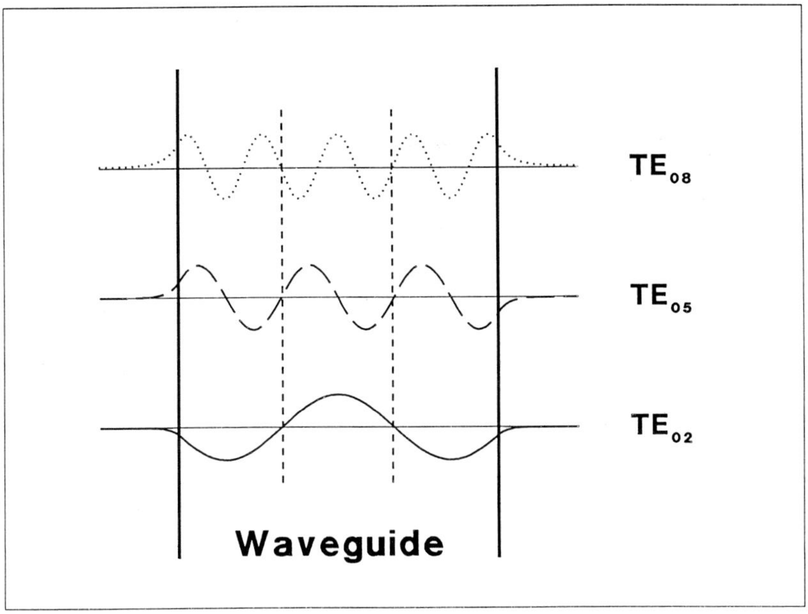

Figure 5.22 The modes of order 2, 5, and 8 in a strongly multimoded MMI section. At 1/3 and 2/3 of the waveguide width (dashed vertical lines) all modes have amplitudes close to zero.

The analysis outlined above provides a basis for the intuitively developed strategy for widening the coupler section and concurrently shifting the access waveguides such as to avoid excitation of the second order mode, as described in the previous section.

5.4.3 MMI-switches

The 3L MMI-coupler configuration can be employed, in principle, as a switch by electro-optically varying its width, for example through carrier injection. From equation 5.16b it can be seen that $\Delta\beta_{01}$ depends inversely quadratic on the waveguide width w. If the waveguide width is reduced by a factor $\sqrt{2}$ the value of $\Delta\beta_{01}$ doubles. Consequently, if the width of a 3L MMI-coupler, operating in the cross state $(\Delta\beta L=3\pi)$ is reduced by a factor $\sqrt{2}$ (i.e. approximately 30%), it will switch to the bar state $(\Delta\beta L=6\pi)$. Figure 5.23 illustrates the corresponding switch configuration.

Because the electro-optically induced changes in refractive index are small, the numerical aperture of the MMI-section will be small, which impedes circuit miniaturization.

It is concluded that MMI-couplers are promising as a process-tolerant alternative to conventional directional couplers and switches. In high-contrast waveguide systems they have the additional advantage of small device dimensions.

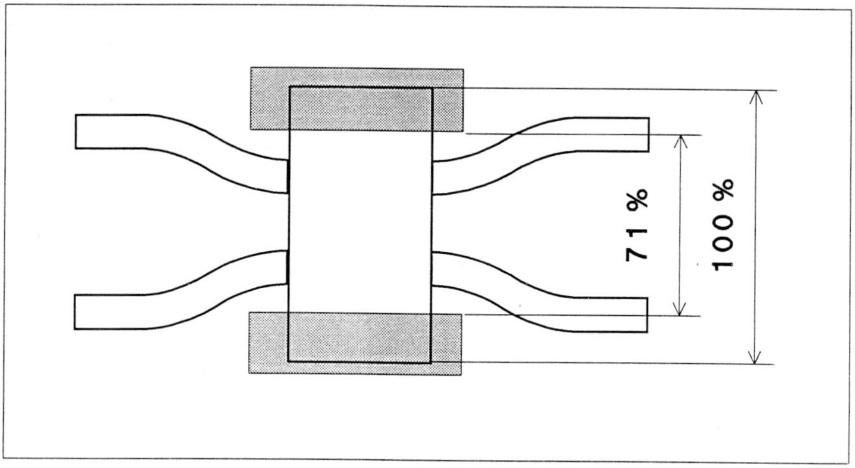

Figure 5.23 The MMI coupler used as a switch.

5.5 Phase-diversity coupling network

5.5.1 The network scheme

Homodyne optical detection converts the optical spectrum to base-band location and therefore imposes very stringent requirements on the spectral purity of both the transmitter and the local-oscillator laser. These requirements can be relaxed by applying phase-diversity detection techniques [16,17].

An elementary coherent detector consists of a 3-dB coupler, in which the received signal and the local-oscillator signal are combined, followed by two photodiodes, where the combined signal is detected and the frequency conversion takes place. It has been shown [16] that pseudo-homodyne detection and cancellation of Relative Intensity Noise (RIN) in the output signals can be achieved simultaneously, without the need of phase-locking the local oscillator, if two output signals with quadrature phase are available. For passive 2x2 ports such as 3-dB hybrids the phase difference between the two output ports after detection differs fundamentally by 180°, so that additional processing is required for arriving at 90° phase difference.

The required quadrature phase relation can be obtained by proper combination of the output signals of 3x3 or 4x4 coupling networks. In planar technology it is difficult to realize a 3x3-port network. Hoffmann et al. [15] reported realization of a 4x4-port network, as depicted in figure 5.24, with electrically controlled couplers and phase shifters in lithium niobate. A first attempt to realize a similar network without electrically tunable components

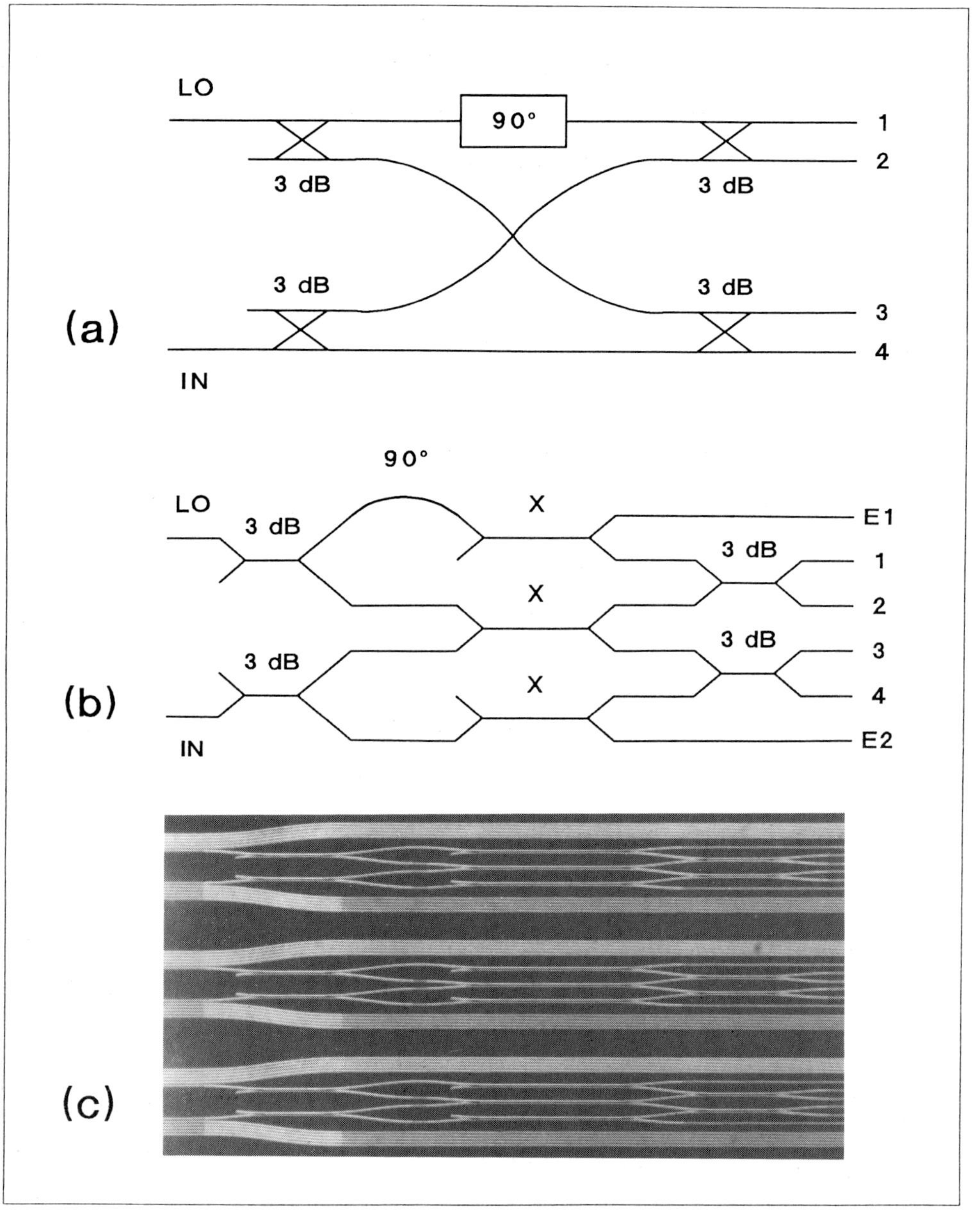

Figure 5.24 4x4-Port phase-diversity network.

a) Schematic diagram
b) Technical implementation
c) Microscope photograph of a realized network.

is reported in this thesis. The project served as a pilot project for a research program[1] which is aimed at realization of a phase diversity network operating at 1550 nm, with less than 2 dB insertion loss and a maximal phase deviation between output ports of ± 20°.

5.5.2 Technical implementation

Figure 5.24b shows how the scheme represented in figure 5.24a can be realized. For the whole circuit to function within specifications the tolerance requirements for the separate components must be strict. Since MMI-couplers were unknown when the presently reported experiment was designed, the TMI-couplers with 2000 μm bends and a 0.5° Y-junction half angle, as reported in section 5.2, were selected as 3-dB couplers. The 90° phase shift was realized by replacing a straight waveguide by a slightly curved one which introduces a differential ¼λ length.

The waveguide crossing was realized as a TMI cross coupler, which is identical to the 3-dB coupler except that it has a longer TMI-section. Cross-couplers were also inserted into the channels parallelling the crossing in order to achieve that phase transfer through these channels is equal to that of the central channel. This is required for proper operation of the network. An additional advantage of this solution is that two extra channels are then available to monitor coupler performance. If the cross couplers function well, the extra channels should carry no signal power. Since the difference between

1) Innovation-directed Research Program (IOP) IC-Technology, subprogram Electro-Optics (funded by the ministry of Economic Affairs).

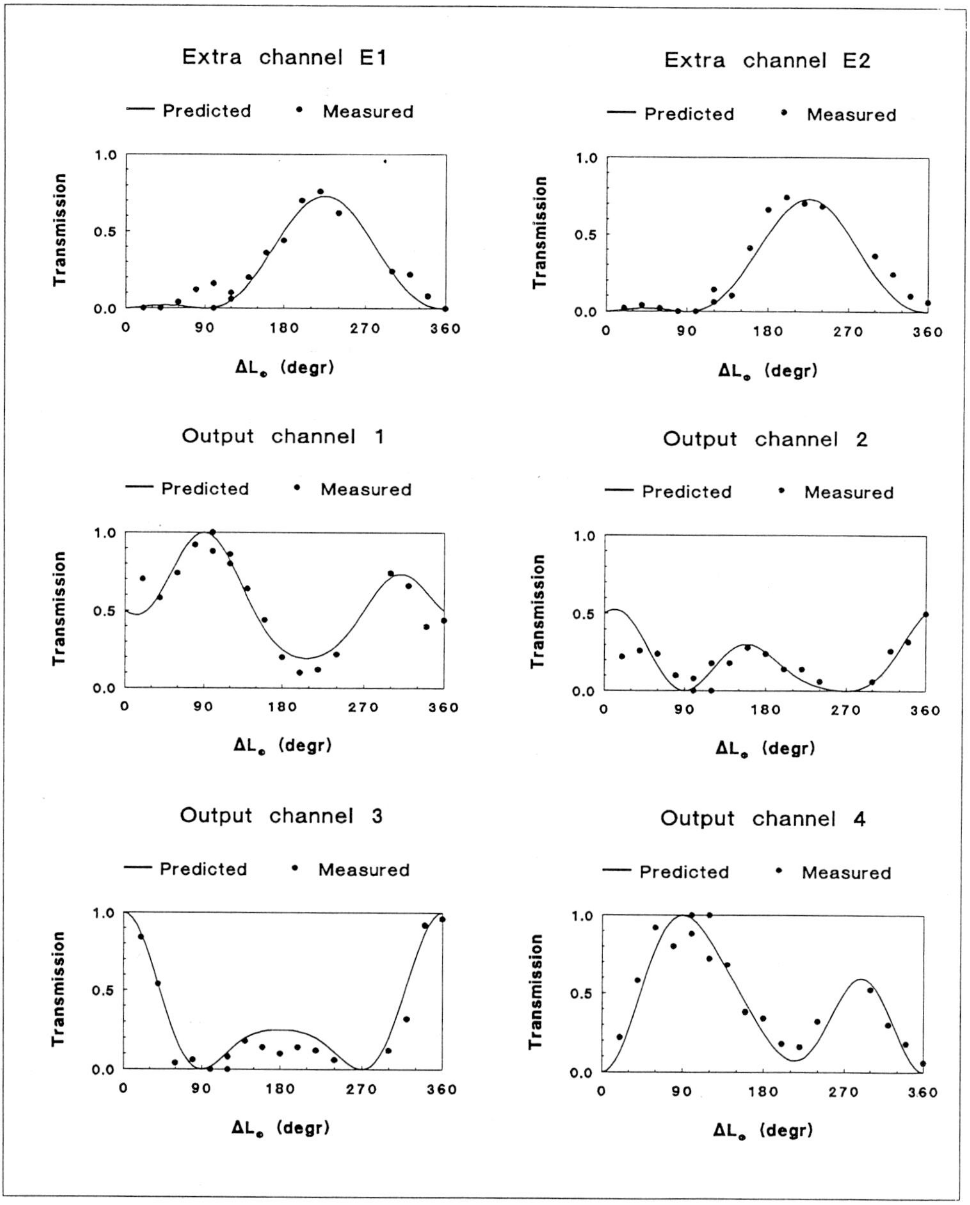

Figure 5.25 *Predicted and measured output intensities of the four output channels of the phase diversity network, and the two extra output channels (as shown in figure 24b), as a function of the additional (normalized) length* ΔL_{Φ} *relative to the length, required for proper operation. The additional length was applied to both the 3-dB and the cross couplers.*

the cross couplers and the 3-dB couplers is restricted to a a short piece of TMI-section, proper functioning of the cross couplers indicates the 3-dB couplers are functioning well too.

The distance between the input channels was chosen to be 125 μm, in order to allow for separate excitation of both inputs through a prism coupler with low cross talk. Output channels were positioned at 50 μm distance for compatibility with the detector diode array used for phase measurements. A series of reference channels were included at both sides of the network to serve as a 100% transmission reference. Figure 5.25c shows some of the realized networks. Circuit dimensions are 5 x 0.25 mm.

5.5.3 Experimental results

A series of phase diversity networks were realized with varying coupler lengths. The cross-couplers were given an additional 48.3 μm length, relative to the 3-dB couplers, to provide for a 90° additional phase difference. Coupler lengths were varied such that the additional normalized length ΔL_{Φ} (defined according to equation 5.13), relative to the values required for proper operation in a phase diversity receiver, ranged from 0° to 360° in steps of 20°. The same value of $\Delta\Phi$ was applied both to the 3-dB and the cross couplers.

Computation of the network response as a function of the additional length ΔL_{Φ} is straightforward. Output intensity measurements were performed as described in chapter 8. Figure 5.25 shows the predicted and measured intensity response for the output channels (as depicted in figure 5.24b) when the two input channels are excited with equal amplitude and phase. This excitation state was experimentally obtained by exciting both input ports simultaneously with a broad input beam, and adjusting the angle between substrate and laser beam for an optimal fit between predicted and measured output intensities, for the network with the lowest intensity in the extra output channels. Due to uncertainty as to the contribution of the Y-junctions, the position of the measured results on the Φ-axis was not known a priori. It was determined by fitting the intensities measured in the extra channels to the predicted response.

The measured output intensities correspond well with the predicted values. Insertion loss of the networks, as determined from figure 5.25a by adding the intensities measured in the six output channels, is smaller than 2 dB for most networks. This is acceptable for many applications.

Phase measurements were carried out too, but did not yield satisfactory results. Improvements are being made both to the circuit and the measurement set-up. An experiment at 1550 nm wavelength with MMI-couplers is presently being prepared by L.B. Soldano of Delft University of Technology and J. van Bennekom of Eindhoven University of Technology.

5.6 Conclusions

Experimental results demonstrate that modified TMI couplers combine small dimensions (submillimeter length) and low insertion loss (<0.5 dB) with good reproducibility figures (phase parameter deviation < ± 10°). Even better results are expected with couplers based on Multi-Mode Interference. With these couplers realization of a small-size phase diversity network with a maximum insertion loss of 2 dB should be feasible.

References

5.1 M. Papuchon, A. Roy, and. D.B. Ostrowski, "Electrically active optical bifurcation: BOA", Applied Physica Letters, Vol. 31, pp. 266-267, 1977.

5.2 R.A Forber and E. Marom, "Symmetric directional coupler switches", IEEE J. on Quantum Electron., Vol. 22, pp. 911-918, 1986.

5.3 A. Neyer and D. Dang, "Guide-wave polarization splitter based on two-mode interference", Procs. Conf. on Integrated and Guided Wave Optics (IGWO'88), 1988, March 28-30, Santa Fe, U.S.A, paper ME3, 1988.

5.4 A. Neyer, "Integrated-optical multichannel wavelength multiplexer for monomode systems", Electron. Lett., Vol. 20, pp. 744-746, 1984.

5.5 F. Rottmann, A. Neyer, W. Mevenkamp, and E. Voges, "Integrated-optic wavelength multiplexers on lithium niobate based on two mode interference", IEEE J. Lightw. Technol., Vol. 6, pp.946-952, 1988.

5.6 C. de Bernardi, S. Morasca, C. Rigo, B. Sordo, A. Stano, I.R. Croston, and T.P. Young, "Wavelength demultiplexer integrated on AlGaInAs/InP for 1.5 μm operation", Electron. Lett., Vol. 25, pp. 1488-1489, 1989.

5.7 Y. Chung, J.C. Yi, S.H. Kim, and S.S. Choi, "Analysis of a tunable multichannel two-mode-interference wavelength division multiplexer/demultiplexer", IEEE J. Lightw. Technol., Vol. 7, pp. 766-777, 1989.

5.8 P.J. Schalkwijk, Planair Phase Diversity netwerk voor een coherente optische ontvanger (Planar Phase Diversity network for a coherent optical receiver), Masters Thesis, Delft University of Technology, Dept. of Electrical Engineering, Laboratory for Telecommunication and Remote Sensing Technology, Delft, The Netherlands, January 1989.

5.9 F.B. Veerman, P.J. Schalkwijk, E.C.M. Pennings, M.K.Smit, and B.H. Verbeek, "A passive 3-dB TMI-coupler with reduced fabrication tolerance", to appear in J. Lightw. Technol.

.F.B. Veerman, Een passieve 3-dB koppelaar met verminderde gevoeligheid voor productietoleranties, Masters Thesis, Delft University of Technology, Dept. of Electrical Engineering, Laboratory for Telecommunication and Remote Sensing Technology, Delft, The Netherlands, September 1990.

5.10 E.C.M. Pennings, "Bends in Optical ridge waveguides", Ph.D Thesis, ISBN 90-9003413-7, Delft University of Technology, Delft, 1990.

5.11 O. Bryngdahl, "Image formation using self-imaging techniques", J. Opt. Soc. Am., Vol. 63, pp. 416-418, 1973.

5.12 R. Ulrich and G. Ankele, "Self-imaging in homogeneous planar optical waveguides", Appl. Phys. Lett., Vol. 27, pp. 337-339.

5.13 R. Ulrich, Verfahren und Vorrichtung zur Strahlenteilung oder Strahlenkopplung unter Verwendung eines Wellenleiters, Auslegeschrift 25 06 272, Deutsches Patentamt, Anmeldetag 14 Februar 1975.

5.14 H.-G. Unger, Planar optical waveguides and fibres (Oxford: Clarendon Press, 1977).

5.15 D. Hoffmann, H. Heidrich, G. Wenke, and R. Langenhorst, "Integrated optics 90° hybrid on $LiNbO_3$ for phase diversity receivers", Electron. Lett., Vol. 24, pp. 1374-1375, 1988.

5.16 W. van Etten, "Coherent optical fibre communication", Tijdschrift van het Nederlands Elektronica- en Radiogenootschap (Journal of the Dutch society of electronic and radio engineers), Vol. 55, pp. 89-97, 1990.

5.17 J. Siuzdak and W. van Etten, "BER evaluation for phase and polarization diversity optical homodyne receivers using noncoherent ASK and DPSK demodulation, J. Lightw. Technol., Vol. 7, pp. 584-599, 1989.

Chapter 6

Optical phased arrays

In this chapter, the phased-array concept is applied to the design of novel planar optical components which combine focusing and dispersive properties with small dimensions. The array performance for spatial separation or combination of different wavelengths (demulti/multiplexers) or polarizations (polarization splitters/combiners) is modeled, and a method is developed for an optimal design of these components. Finally, experimental results on a four-channel wavelength demultiplexer and a polarization splitter are presented.

6.1 Introduction

Focusing and dispersive elements play an important role in integrated optical circuits. Components which combine both functions are of special interest for the spatial separation of signals with different wavelengths, as occurring in wavelength demultiplexers and wavelength filters. They may also be used for polarization splitting if the waveguides employed exhibit polarization dispersion.

Planar lenses have been reported in many kinds and dimensions [1-3]. The dimensions of most lens types are large (millimeters to centimeters), which make them unattractive from a viewpoint of circuit miniaturization. Fresnel lenses [3] and hyperbolical index-contrast lenses [4] were reported with sub-millimeter dimensions. To realize large numerical apertures, however, both lens types require high index contrasts.

All the lens types mentioned above are, in principle, nondispersive, so that they have to be combined with a dispersive element for wavelength separation purposes. The only small sized component combining focusing and dispersive properties which can be realized with low optical contrasts is the curved planar grating [3]. A disadvantage of such a grating is that it cannot be realized using conventional optical lithography but requires submicron lithography (holographic or electron beam).

In 1983 we developed a new concept for a focussing and dispersive component with properties comparable to a grating. It can be realized simultaneously with the waveguide pattern employing conventional optical lithography, thus avoiding the additional process steps involved in applying holographic or direct-write electron-beam lithography. The concept is based on a ***phased array*** of bent optical waveguides, the radius of which is used to control the length of the waveguides and, consequently, the phase transfer. It was not until 1986 that our aluminum oxide ridge-guide technology was sufficiently developed to allow a first prototype to be realized [5]. In 1988 we realized a polarization splitter [6,7] and a polarization-independent wavelength demultiplexer [8,9] based on this concept, both of which combined state-of-the-art performance for planar components with the smallest device size so far reported. The components were realized for operation at short wavelengths (633 and 780 nm) because of the availability of accurate measurement equipment for these wavelengths. Recently Takahashi et al. [10,11] of NTT reported realization of a 1.3 μm wavelength demultiplexer using the above principle.

In the following sections analysis and design of focussing and dispersive components based on an optical phased array is described and experimental results demonstrating the technological feasibility of the concept are given.

6.2 Basic principles

If a broad parallel beam is impinging on an array of concentric planar optical waveguides, as shown in figure 6.1, part of the incident power will couple into the waveguides and the other part will propagate straight forward or be scattered.

The light coupled into the waveguides will propagate to the output aperture and arrive there with a phase distribution

$$\Phi_i = \beta\psi R_i + \Phi_o, \qquad (1)$$

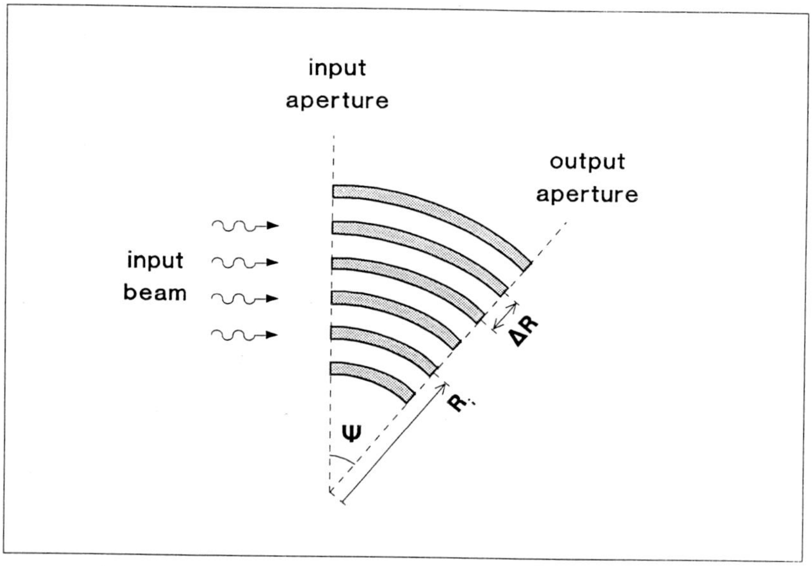

Figure 6.1 Phased array geometry.

in which β is the propagation constant of the fundamental waveguide mode, ψ is the sector angle of the concentric array, and R_i is the bending radius of the i-th waveguide. The phase Φ_o at the input aperture is not relevant for the following analysis. If we choose the concentric waveguides such that:

$$\beta\psi R_i = \beta\Psi R_1 + 2\pi m\,(i\text{-}1), \tag{2}$$

then the phase distribution at the end face of the array will be uniform and a plane wave perpendicular to the end face will be radiated. The integer m is the order of this beam.

The phase distribution need not necessarily be uniform; each required distribution is realized equally well by computing the R_i from:

$$\beta\psi R_i = \beta\Psi R_1 + \Phi(R_i) + 2\pi m\,(i\text{-}1), \tag{3}$$

in which $\Phi\,(R)$ is the required phase distribution. A quadratic distribution will yield focusing operation.

The phased array is principally dispersive. The phase at the end of the waveguide channels is proportional to their length. Because this length is linear in R a frequency shift df will cause a phase shift which is linear in R_i, thus tilting the wave front without affecting its curvature. Consequently, the position of the focal spot in the image plane will shift linearly with df while the focal length remains unaffected.

The dispersion is easily inferred from figure 6.2. The direction θ_m of the outgoing wavefront corresponding to the *m-th* order follows as the arcsine of the path $\Delta\Phi/\beta$ divided by the waveguide spacing ΔR, as indicated in the figure. The phase difference $\Delta\Phi$ contains a multiple m times 2π; this multiple has to be subtracted in order to find the propagation angle θ_m of the outgoing beam:

$$\theta_m = arcsin\left\{\frac{(\Delta\Phi - 2\pi m)\,/\,\beta_f}{\Delta R}\right\} = arcsin\left\{\frac{r\,(\Delta l - m\lambda_g)}{\Delta R}\right\}, \tag{4}$$

in which Δl is the difference in length between adjacent waveguides, β_f is the (two-dimensional) plane-wave propagation constant in the film which guides

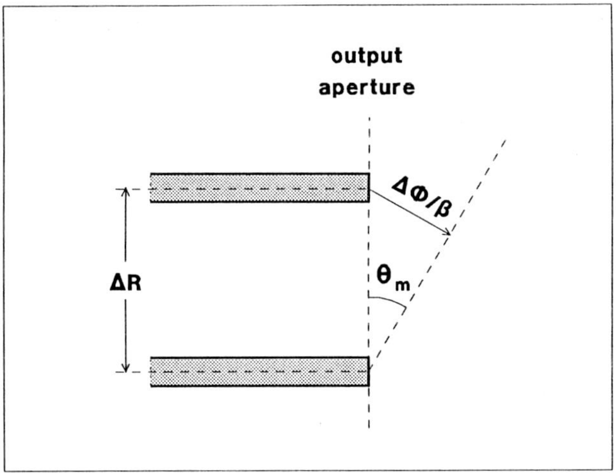

Figure 6.2 Relation between the wave front tilting angle θ and the phase difference ΔΦ between adjacent waveguides.

the light on both sides of the array, λ_g is the wavelength measured within the waveguide, and r is a constant describing the refraction at the end of the array:

$$r = \frac{\beta_g}{\beta_f} = \frac{\lambda_f}{\lambda_g}. \qquad (5)$$

For low-contrast waveguides the value of r will be close to one. The precise distance ΔR_i between adjacent waveguides follows from the values R_i which satisfy equation 6.3. It may be different from one waveguide to another, dependent on the required phase distribution. If the order m of the beam considered is not to low the differences will be small, however. By the distance ΔR, *without index i,* we will understand the average value of ΔR_i.

Radiation will occur into those orders for which the argument of the arcsin is within -1 and +1. The dispersion is derived by differentiating equation 6.4 with respect to λ_g. Using the identity $\Psi\Delta R = m\lambda_g$ which holds for orders with $\theta_m \ll \frac{1}{2}\pi$, we find:

$$\frac{d\theta_m}{d\lambda_g} \simeq -\frac{mr}{\Delta R} = -r\frac{\Psi}{\lambda_g}, \tag{6a}$$

or

$$\Delta\theta \simeq -r\frac{\Delta\lambda}{\lambda_c}\Psi. \tag{6b}$$

in which λ_c is the central wavelength, for which $\theta_m = 0$. From the latter formula it is seen that a wavelength shift of 1% will cause an angular dispersion of 0.01Ψ; with $\Psi=\pi$ this is roughly 2°.

A set of five arrays, operating at 633 nm wavelength, was realized with Al_2O_3 ridge guides on a silicon substrate, as described in chapters 2 and 3. Figure 6.3 shows a photograph of the arrays (the white areas). We chose a number of 31 concentric waveguides with a width of 3 μm at an average distance[1] ΔR of 6 μm, with R_i varying between 910 and 1090 μm. The concentric section sector angle Ψ was chosen to be 30°.

---

1) The array was designed to fit the required phase distribution $\Phi(R)$ ***modulo 2π*** in order to avoid the waveguide spacing varying strongly across the aperture. If the order is 10, for example, phase variations between $-\pi$ and $+\pi$ can be obtained by varying the mutual distance $\Delta R_i = R_i - R_{i-1}$ between plus or minus 5%. Employing this approach the array is only slightly different from an equidistant array.

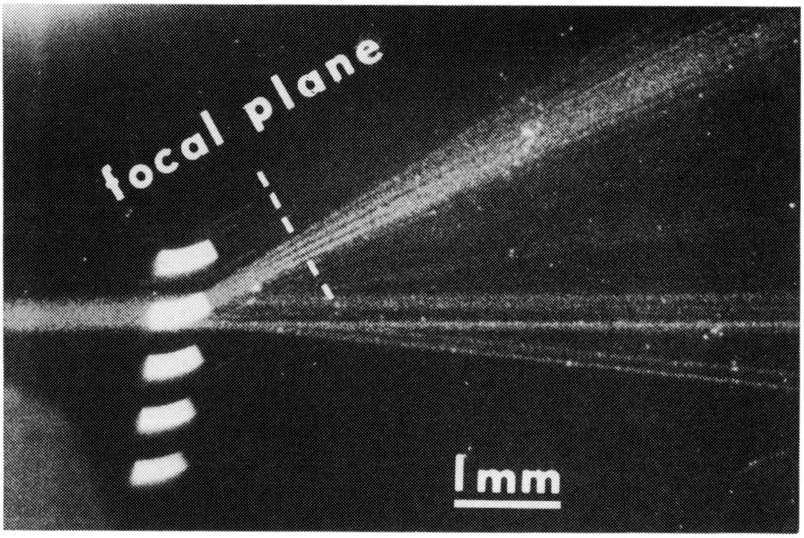

Figure 6.3 *Focusing phased array showing multiple orders. The photograph shows five arrays, of which the second one from the top is excited with a He-Ne laser beam.*

The photograph shows a number of focused beams, corresponding to different orders, emerging from the second array, which is excited with a Gaussian He-Ne beam filling the array input aperture (width 180 μm). The dashed line represents the design position for the focal plane (f = 1 mm). The spatial separation of the orders in the focal plane compares well with the expected value[1]: 67 μm.

The occurrence of multiple beams prohibits efficient coupling of the outgoing beam into a single waveguide. To avoid the occurrence of multiple orders,

---

1) The angular spacing between the beams is expected to be $(\lambda_o /N)/\Delta R \simeq 67.10^{-3}$ (N is the effective index: $N = \beta/k_o \simeq 1.57$; k_o and λ_o are the vacuum wavenumber and wavelength, respectively). With a focal length $f = 1000\ \mu m$ the spacing of the orders in the focal plane should amount to $\simeq 67\ \mu m$.

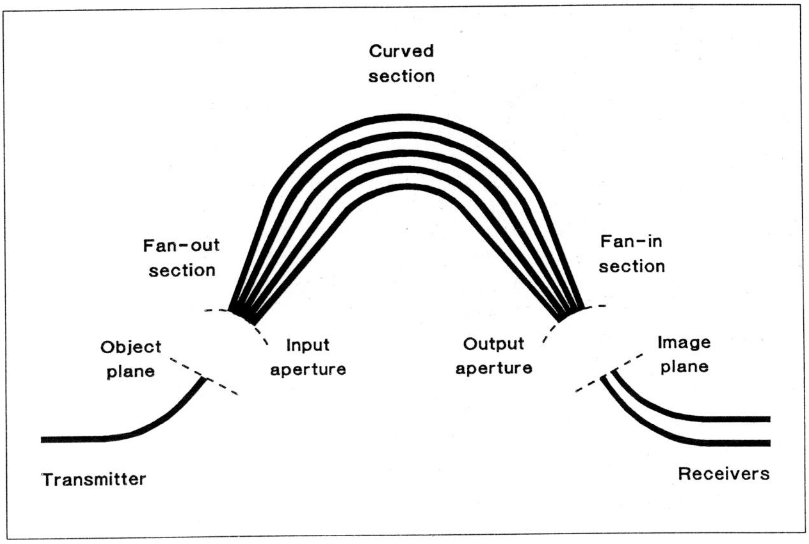

Figure 6.4 *Focusing and dispersive phased array with fan-out and fan-in coupling sections.*

and simultaneously reduce coupling losses, the array should be provided with fan-out and fan-in coupling sections as shown in figure 6.4. By bringing the waveguide ends closer together the angular distance between the different orders will increase and power coupled into beams alongside of the waveguide axis will be reduced. If the waveguides touch each other most of the power will be coupled into one single order and efficient coupling into a single waveguide becomes feasible.

In the following sections design and analysis of a wavelength demultiplexer and a polarization splitter, i.e. a device with one transmitter waveguide and more than one receiver waveguide, will be discussed. The component being reciprocal, design and analysis of a multiplexer or a polarization combiner are similar; the only difference is the transmission direction.

6.3 Array geometry

In the design as illustrated in figure 6.4, the optical lengths of the array waveguides determine the dispersive and focusing properties of the component. The performance is, to a large extent, independent of the precise shape of the waveguide pattern, and different approaches may be adopted for realizing the required phase transfer. Figure 6.5 illustrates four of them.

In the first experiments we applied the geometry depicted in figure 6.5a, which consists of a concentric section provided with fan-in and fan-out sections which are connected smoothly with the concentric section by a double-bend adapter section. According to this concept, which is described in

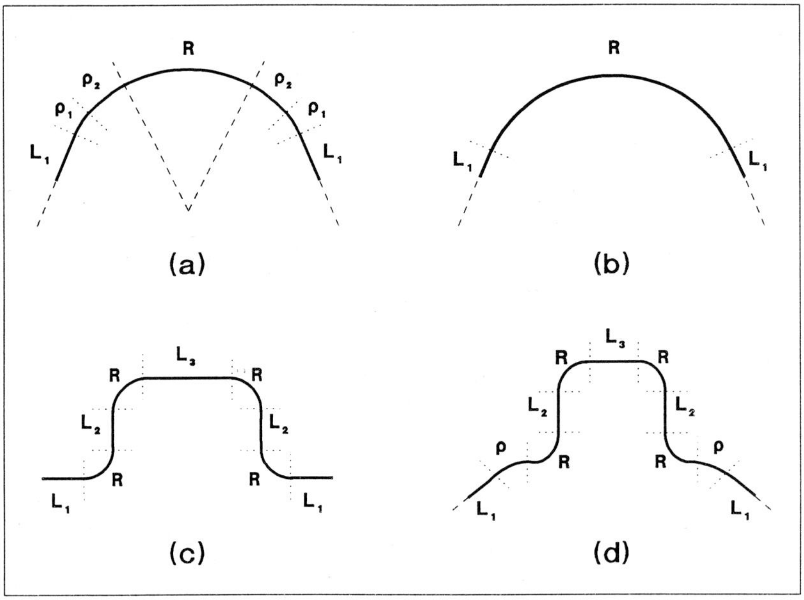

Figure 6.5 Possible geometries of a phased-array channel.

in the appendix at the end of this chapter, the dependence of the channel length on the rank number contains quadratic and higher-order components, which are introduced by the adapter section. The quadratic terms cause the wavefront curvature and, consequently, the position of the focal point to change if the wavelength changes (chromatic aberration). The effect can be compensated by adapting the positions of the receiver waveguides perpendicular to the image plane. Due to shadowing effects cross-talk performance degrades, however, as will be discussed in more detail in section 6.7.

Takahashi et al. [10,11] employed the structures depicted in figures 6.5c and 6.5d. These structures are free of chromatic aberration and very simple from a design point of view. The latter structure, which is necessary if low insertion coupling loss is to be achieved, suffers from a large number of junctions between straight and curved waveguides, which introduce mode conversion and coupling loss if short bends are employed.

In the present thesis the geometry shown in figure 6.5b, which contains a minimal number of waveguide sections and junctions, will be further elaborated. According to this concept each array waveguide channel is composed of a curved section enclosed by two straight ones. Figure 6.6a shows the i-th element of the array in more detail.

For the array to be focusing, the optical lengths of all elements have to be equal except for an integer number *m* of wavelengths, which is the order of the focused beam. Each element path is described by its starting angle α_i ,

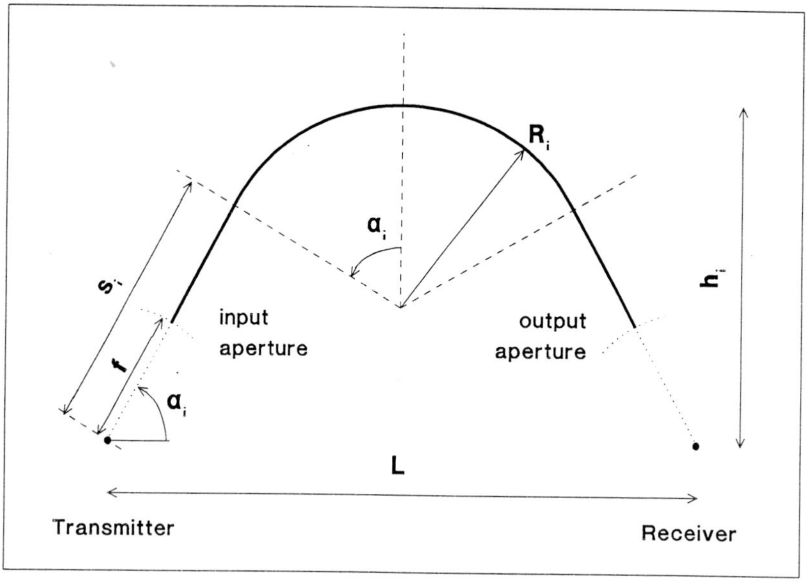

Figure 6.6a *Geometry of the i-th channel of the phased array.*

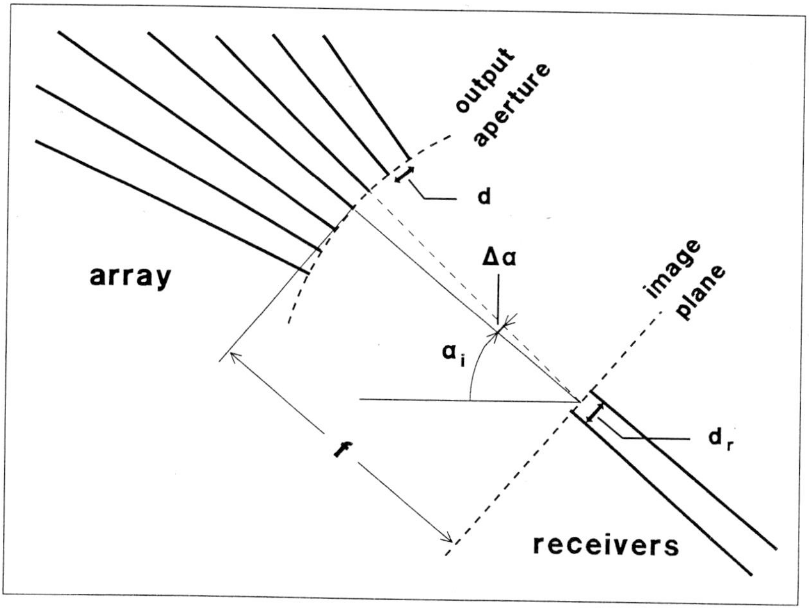

Figure 6.6b *Geometry of the receiver side of the phased array. Note that the angle* α_i *is measured clockwise at the receiver and counterclockwise at the transmitter.*

the radius of curvature R_i , and the straight section length s_i , which is measured from the focal points to the curved section. These parameters follow from the array parameters according to[1] (see figures 6.6a and 6.6b):

$$\alpha_i = \alpha_1 + (i\text{-}1)\,\Delta\alpha, \qquad i = 1, n_a, \tag{7}$$

$$s_i = \tfrac{1}{2}\left(l_i - \frac{L}{sinc\ \alpha_i}\right) / \left(1 - \frac{cos\ \alpha_i}{sinc\ \alpha_i}\right), \tag{8}$$

and

$$R_i = \frac{\tfrac{1}{2}L - s\ cos\ \alpha_i}{sin\ \alpha_i}, \tag{9}$$

in which l_i is the length of the i-th element:

$$l_i = l_1 + m\,(i\text{-}1)\,\lambda_c, \tag{10}$$

and λ_c is the central design wavelength[2] measured within the waveguide. L is

---

1) The distance L between transmitter and receiver can be written as:

$$L = 2s_i\, cos\, \alpha_i + 2R_i\, sin\, \alpha_i\,.$$

The length of the path l_i from transmitter to receiver, measured along the array channel, follows as:

$$l_i = 2s_i + 2\alpha_i R_i\,.$$

The relations 6.8 and 6.9 follow directly from the above relations.

2) The central design wavelength λ_c is the wavelength (measured within the waveguide) for which the array has a uniform phase transfer; i.e. the source field is imaged in the center of the focal plane (θ_m=0).

the distance between the focal points at the transmitter and the receiver side:

$$L = 2\,(s_i \cos \alpha_i + R_i \sin \alpha_i), \tag{11}$$

and is the same for all elements. The array ends in both sides in a free propagation region (a planar film) with length f. The length f of this region depends on the angular spacing $\Delta\alpha$ of the array elements and their mutual distance d in the array output aperture (i.e. the junction between the array and the free propagation region) according to

$$f = \frac{\tfrac{1}{2}d}{\arctan(\tfrac{1}{2}\Delta\alpha)} \simeq d/\Delta\alpha. \tag{12}$$

Equations 6.7 to 6.12, together with the length f of the free propagation region, fully define the i-th element. The free propagation length f is minimal if d equals the waveguide width w. In practice f will be larger due to partial filling of the inter-waveguide gap as a result of the finite resolution of the lithographic process. If uncertainty about the actual value of f is to be avoided it should be chosen such that the gap width at the junction is larger than the resolution of the lithographic process (with optical lithography between 0.5 and 1 µm).

If the free propagation region is masked and processed simultaneously with the waveguides, the discontinuity at the end of the waveguides will be minimal. In ridge-type waveguides this solution has the additional advantage that the unetched parts of a film have a lower attenuation than the etched ones.

6.4 Analysis of focusing and dispersive properties

Analysis of wavelength demultiplexers and polarization splitters differs in that the wavelength is a continuous variable and the polarization a discrete one. Yet, a general description, applying to both device types, can be obtained by characterizing the channels by their wavelengths $\lambda_{gi}=\lambda_i/N_i$ inside the array waveguides (in which λ_i is the vacuum wavelength corresponding to the i-th wavelength channel), because these wavelengths, which respond both to the vacuum wavelength and the polarization, fully determine the transfer properties of the array. For a demultiplexer the effective indices N_i will be approximately equal for all channels, but the wavelengths are different. For a polarization splitter the vacuum wavelengths λ_i are identical, but the effective indices are different. Both effects lead to different values for the internal wavelengths λ_{gi}. For the analysis the origin of these differences: wavelength or polarization, is not relevant.

In the following subsections the dispersion, insertion loss, and cross-talk properties of the (generalized) device and the polarization dependence of wavelength demultiplexers will be analyzed.

6.4.1 Dispersion

For the radiation angle of an array as described in the previous section, we find, analogous to the derivation of equation 6.5:

$$\theta_m = \arcsin\left[\frac{(\Delta\phi - 2\pi m)/\beta_f}{d}\right] = \arcsin\left[\frac{r(\Delta l - m\lambda_g)}{d}\right], \tag{13}$$

in which d is the waveguide spacing in the array aperture, and the other parameters are as defined in equations 6.4 and 6.5. Differentiation of equation 6.13 with respect to λ_g yields:

$$\frac{d\theta_m}{d\lambda_g} \simeq -r\frac{m}{d}, \tag{14}$$

From this equation the dispersive displacement $dy/d\lambda_g$ of the focal spot in the image plane (y is the coordinate along the image plane) is found by multiplication with the image distance f and substitution of equation 6.12:

$$\frac{dy}{d\lambda_g} = f\frac{d\theta_m}{d\lambda_g} \simeq -\frac{mr}{\Delta\alpha}, \tag{15a}$$

or, in terms of the relative wavelength dispersion δy

$$\delta y = \frac{dy}{\delta\lambda} = -r\frac{m\lambda_c}{\Delta\alpha}, \tag{15b}$$

in which $\delta\lambda$ denotes the relative wavelength variation

$$\delta\lambda = \frac{\Delta\lambda_g}{\lambda_c}, \tag{15c}$$

and λ_c is the central design wavelength measured within the waveguide. The dispersive displacement in the image plane appears to be independent of the length f of the free propagation region. The increase in the length of this region, which may result from partial filling of the gap between the waveguides in the fan-in and fan-out sections, is thus expected to have little or no effect on the dispersion properties of the array.

Deviations of the effective mode index from the design value, as may be due to process deviations in film thicknesses, film refractive indices, lateral index contrasts or waveguide widths, will affect the propagation constant β_g and have the same effect as a wavelength shift of the source of the same relative magnitude.

6.4.2 Insertion Loss

A. Diffraction loss

The field in the image plane at the receiver side is the product of the array pattern and the far field of the array waveguides, as depicted in figure 6.7b. If the focal spot moves away from the center of the waveguide far field, its intensity decreases according to:

$$I(\theta) = I_o\, e^{-2\theta^2/\theta_o^2} \tag{16}$$

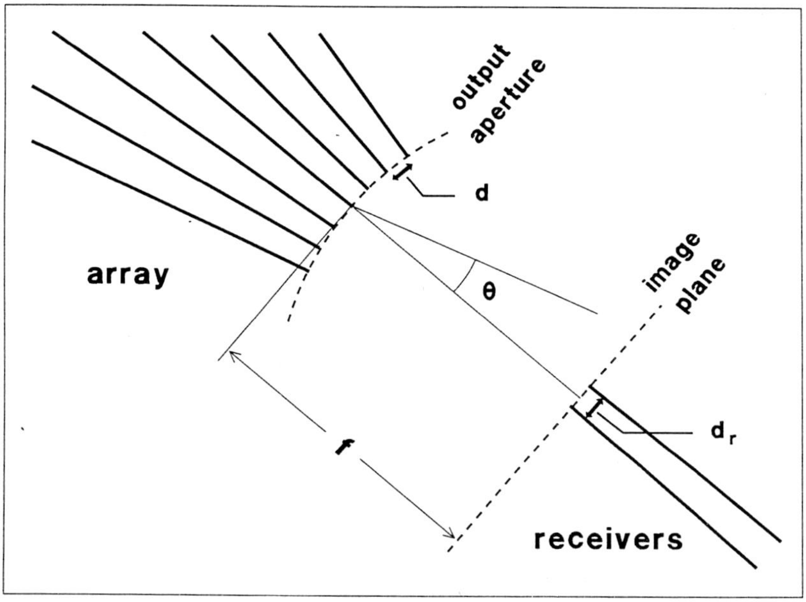

Figure 6.7a *Geometry of the receiver side of the phased array.*

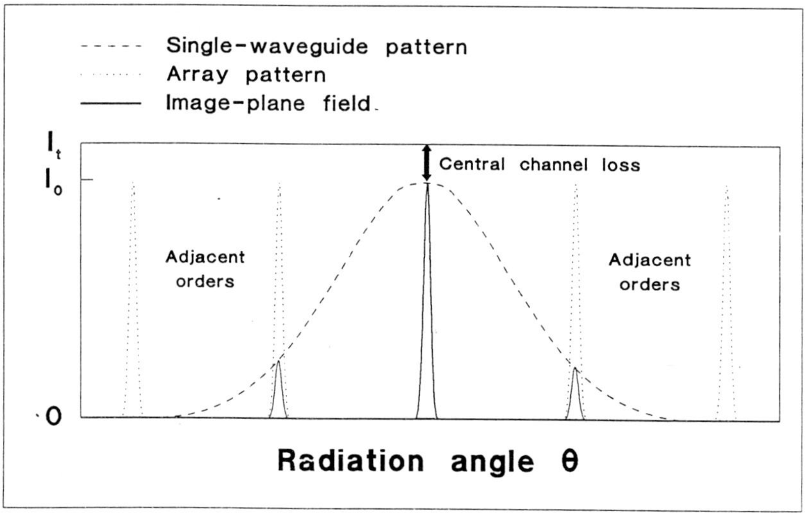

Figure 6.7b *Focal field pattern of the phased array showing the mechanism of diffraction losses.*

In this formula θ_o is the equivalent Gaussian width (see section 3.2.1B) of the waveguide far field:

$$\theta_o = \frac{\lambda_f}{\pi w_o} = \frac{\lambda_f}{w_e \sqrt{2\pi}}, \qquad (17)$$

w_e is the effective width of the waveguide mode (equation 3.11), and λ_f is the wavelength within the free propagation region. The diffraction loss L_{di} of the i-th wavelength channel is the sum of the angular insertion loss $L_{\theta i}$ and the insertion loss L_o of the (virtual[1]) central channel ($\theta = 0$, see figures 6.7a and 6.7b):

$$L_{di} = L_{\theta i} + L_o, \qquad \text{(dB)} \qquad (18)$$

with

$$L_{\theta i} \simeq 10 \log_{10} e^{-2\theta_i^2/\theta_o^2}. \qquad \text{(dB)} \qquad (19)$$

The angle θ_i for the i-th receiver follows from its position y_i in the image plane through

$$\theta_i \simeq y_i / f. \qquad (20)$$

1) In the case of an even number of receiver waveguides there will be no waveguide located at $\theta = 0$. The central channel loss L_o then refers to the loss of a virtual channel at $\theta = 0$.

Angular insertion loss for the outer channels may vary from a few tenths of a decibel to a few decibels, depending on the design. The loss L_o can be estimated as follows. Most of it will be caused by radiation into the adjacent orders of the array pattern. In a Gaussian approximation these orders will be suppressed by a factor $exp\,(-2\Theta^2_{per}/\theta^2_o)$, in which Θ_{per} is the period of the array pattern:

$$\Theta_{per} = arcsin\,(\frac{\lambda_f}{d}) \simeq \frac{\lambda_f}{d}, \tag{21}$$

and d is the waveguide spacing. Power will be lost mainly into the two adjacent orders of the main beam, and the same loss will occur at the transmitter side for reasons of reciprocity. The diffraction loss L_o (at $\theta=0$) thus follows as:

$$L_o = -\,10\,log_{10}\,(1 - 4\,e^{-2\Theta^2_{per}/\theta^2_o}) \simeq 17\,e^{-2\Theta^2_{per}/\theta^2_o}, \tag{22a}$$

or

$$L_o \simeq 17\,e^{-4\pi w_e^2/d^2} \quad \text{(dB)}. \tag{22b}$$

For $w_e/d \simeq ½$, a representative value, estimated loss is 0.7 dB.

B. Spillover

Due to the finite width of the array aperture, part of the beam diverging from the transmitter waveguide will spoil along the array, as shown in figure 6.8, and be lost. If the array half angle θ_{max} is chosen twice the equivalent Gaussian width of the divergent beam the predicted spillover will be less than 0.01%, corresponding with a transmission loss less than 0.001 dB. Although the Gaussian approximation will probably be too optimistic in predicting the side lobe level of the radiation pattern, it is clear that power loss due to spillover will be negligibly small through a wide range of design parameters.

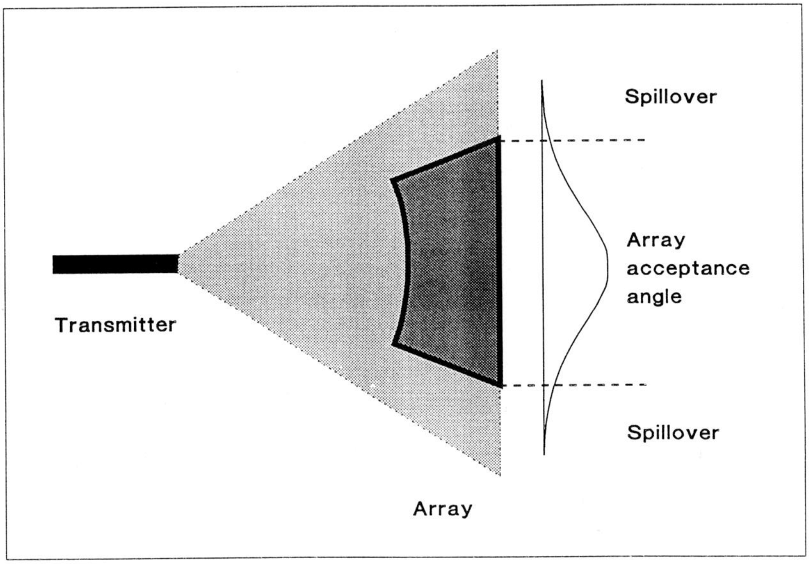

Figure 6.8 *Spill-over due to a finite array aperture.*

C. Other sources of loss

If the array is designed such that the radiation loss in the bend sections is negligible, the loss due to optical waveguide attenuation will be proportional to the size of the array and, more specifically, to the length of the propagation path through the component.

For low contrasts the reflection loss at discontinuities at both sides of the free propagation regions will be small, especially if the transverse waveguide geometry within the waveguides and the free propagation regions are chosen to be identical. This loss may be roughly estimated from the Fresnel reflection coefficient at the junction as computed from the effective indices on both sides. For a relative lateral contrast of 0.05 and a lateral V-parameter of 3, the return loss per junction will be in the order of 40 dB. The corresponding transmission loss is negligible.

6.4.3 Channel cross talk

A. Coupling at the receivers

In the ideal case in which the modal field of the transmitter is imaged onto the image plane without distortion there is still some coupling through the tails of the modal field distribution which will not only excite the intended receiver waveguide, but also the adjacent one(s). The problem is illustrated in figure 6.9.

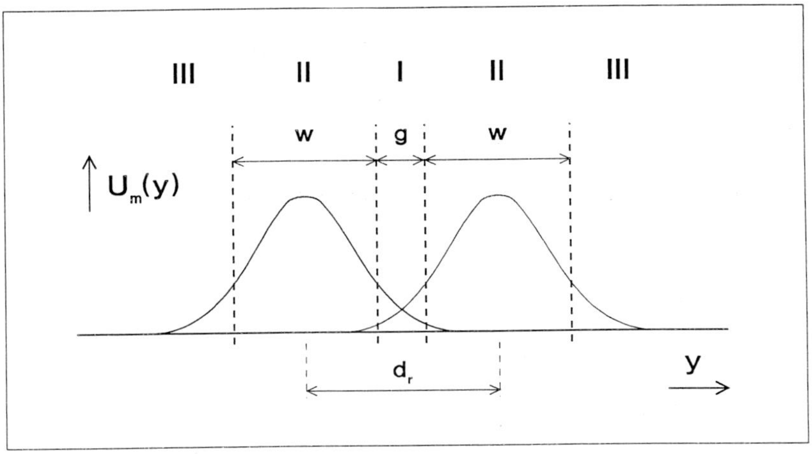

Figure 6.9 Coupling of the focal field to an adjacent receiver channel.

The cross talk due to this phenomenon can be computed from the overlap between the focal field and the modal field of the adjacent waveguide:

$$\eta_c = \frac{\{ \int U_m(y) \;\; U_m(y - d_r) \;\; dy \}}{\{ \int U_m^2(y) \;\; dy \}^2}, \tag{23}$$

in which $U_m(y)$ represents the modal field of the receiver waveguide, and the focal field is assumed to be a perfect image of modal field U_m of the transmitter.

It is seen from figure 6.9 that the overlap integral can be split into three different regions. In each of these regions an analytical expression for the integrand is easily derived from the formulae for the mode-profile (see, for example, Unger [17]). Analytical integration leads to the following expression for the channel cross talk:

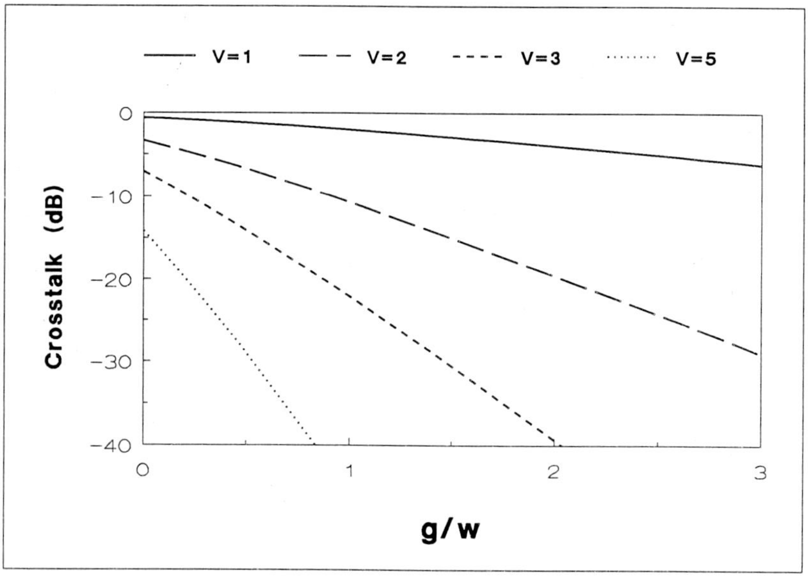

Figure 6.10 Channel cross talk as a function of the relative waveguide spacing.

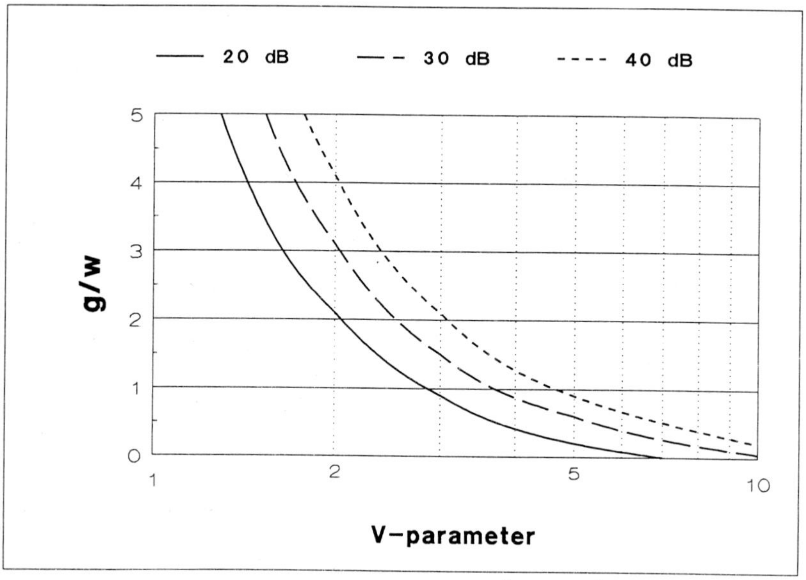

Figure 6.11 Relative waveguide spacing required for -20, -30, and -40 dB cross-talk levels.

$$\eta_c = \left[\frac{2(1-b)\,e^{-vg/w}}{1+2/v}\,(g/w + 4v/V^2 + e^{-v}/v)\right]^2 \tag{24}$$

in which g/w is the ratio of the inter-waveguide gap $g\,(=d-w)$ and the waveguide width w. The three terms in the right-hand factor correspond to the regions *I, II*, and *III* respectively. Figure 6.10 shows the wavelength channel cross talk as a function of this ratio for different values of the V-parameter of the transmitter and the receiver waveguides. Figure 6.11 shows the values of g/w, which are required for obtaining 20, 30, and 40 dB isolation, as a function of the waveguide V-parameter.

B. Interference with adjacent orders

Severe cross talk will occur if the the angular spacing between adjacent orders of the array, as shown in figure 6.7b, equals the angular spacing between the outer receivers. If this is the case it is not possible to excite only one outer receiver; the other one will simultaneously be excited by the adjacent order. If the receivers are positioned symmetrically with respect to the array axis ($\theta=0$, figure 6.7a) both will receive an equal amount of power which means that the cross talk level equals the signal level! The problem should be circumvented by choosing a sufficiently large angular period $\Theta_{per} = \lambda_f/d$ for the array.

This interference phenomenon is closely related to the diffraction loss of the outer channels, as may be seen from a power balance. Full interference will occur for the outer wavelength channel if an adjacent order is imaged onto the opposite outer receiver channel. If this happens, the total diffraction loss L_d

(equation 6.18) of the outer channels is minimally 3 dB, because power is equally divided among these channels. By applying the modus tollens[1] it is concluded that no full interference can occur if $L_d < 3\ dB$ for the outer channels. Some margin is necessary, of course, to prevent coupling through the tails of an adjacent order. It is clear, however, that for low-loss designs this requirement will be met automatically.

C. Truncation of the array aperture field

A third source of cross talk is associated with the side lobes of the focal field pattern. These are the results of truncation of the field at the array aperture as a result of its finite width. This phenomenon is the reciprocal of spillover as discussed in a previous subsection (6.4.2B).

If the truncated aperture illumination is considered to be the sum of the ideal illumination and the inverse of the truncated field, it can easily be seen that the power in the truncated field equals the spillover. In the worst case all of this power is coupled into two adjacent channels (for symmetry reasons), so that the channel cross talk maximally equals half the spillover. For a spillover of 0.01% we thus expect a maximal cross talk of -43 dB.

---

1) Modus tollens: If A implies B, then ***not B*** implies ***not A***.

D. Coupling between the array waveguides

In the fan-out and fan-in sections the waveguides come so close to each other that coupling will occur. To estimate the magnitude of the coupling effects we will consider the coupling between adjacent array waveguides by applying a weak coupling approximation. In this approach the effect of coupling on the signal amplitude in the i-th waveguide is described by[1]:

$$\frac{d}{dz} A_i(z) = jc\ \{A_{i-1}(z) + A_{i+1}(z)\}, \tag{25}$$

in which $A_j(z)$ is the complex amplitude of the modal field $A_j(z)\, exp(-j\beta z)$ in waveguide *j*, and *c* is the coupling coefficient [17][2]:

$$c = C\, e^{-vg/w} \qquad with \qquad C = \frac{2 v u^2}{\beta w^2 V^2 (1 + 2/v)}. \tag{26}$$

The parameters *u, v,* and *V* are the normalized parameters as discussed in chapter 3. In the fan-out and fan-in sections the inter-waveguide gap *g* depends linearly on the longitudinal coordinate *z*:

---

1) Coupling between waveguide *i* and its two neighbors, *i-1* and *i+1*, is described by the following equation [17]:

$$\frac{d}{dz} a_i(z) = -j\beta\, a_i(z) - jc\, \{a_{i-1}(z) + a_{i+1}(z)\}$$

Substitution of $a_m(z) = A_m(z)\, exp(-j\beta z)$, $m=i-1,i,i+1$, in which $A_m(z)$ is a slowly varying function of *z*, and elimination of the propagation factor $exp(-j\beta z)$ yields equation 6.25.

2) In the formula as given by Unger [17] the term 2v in the denominator of equation 3.162 should be 2/v.

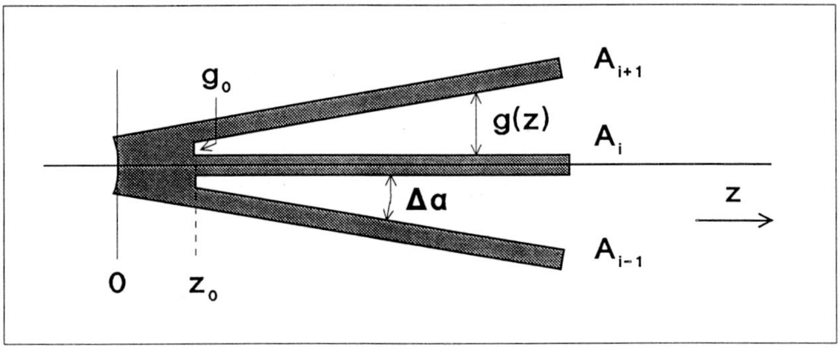

Figure 6.12 Coupling geometry.

$$g(z) = z \, tg \, \Delta\alpha \simeq z \, \Delta\alpha \, , \tag{27}$$

in which $\Delta\alpha$ is the angular increment between adjacent array channels, as shown in figures 6.6a and 6.12. If we are interested in small perturbations of the amplitudes $A_i(z)$ we may consider $A_{i-1}(z)$ and $A_{i+1}(z)$ to be constant in the right hand side of equation 6.25. The total effect ΔA_i of the coupling on the amplitude $A_i(z)$ of the i-th channel is then found by integrating $c(z)$ through the coupling region:

$$\Delta A_i = -j\,(A_{i-1} + A_{i+1}) \int c(z)dz \, . \tag{28}$$

Equation 6.28 is valid as long as $\Delta A_i \ll A_i$. The integral is easily solved analytically:

$$\int_{z_o}^{\infty} c(z)dz = \frac{\lambda_g}{w} \frac{(1-b)}{(1+2/v)\,\theta\pi} \, e^{c g_o / w} \tag{29}$$

with $z_o = g_o/\Delta\alpha$. Figure 6.13 shows the value of $(w/\lambda)\int c(z)dz$ as a function of the gap filling ratio g/w (no filling corresponds to $g/w=0$) for a waveguide with $V=3$, and different values of the inter-waveguide angle $\Delta\alpha$. For $w/\lambda_g \simeq 5$, $\Delta\alpha \simeq 0.02$, and $g/w \simeq 0.5$, which is a realistic value for narrow waveguides, the integral has a value of 0.2.

The effect of the integral in equation 6.27 on the distortion ΔA_i is considerably reduced, however, through multiplication with the sum of A_{i-1} and A_{i+1}. If the amplitude variation from one channel to another is linear, this sum equals $2A_i$ for the whole array (except for the outer elements) and there will be no coupling distortion at all. Only quadratic variations will distort the transfer. If the array encompasses 20 channels and its aperture covers 4 times the equivalent Gaussian width of the aperture field distribution, then the

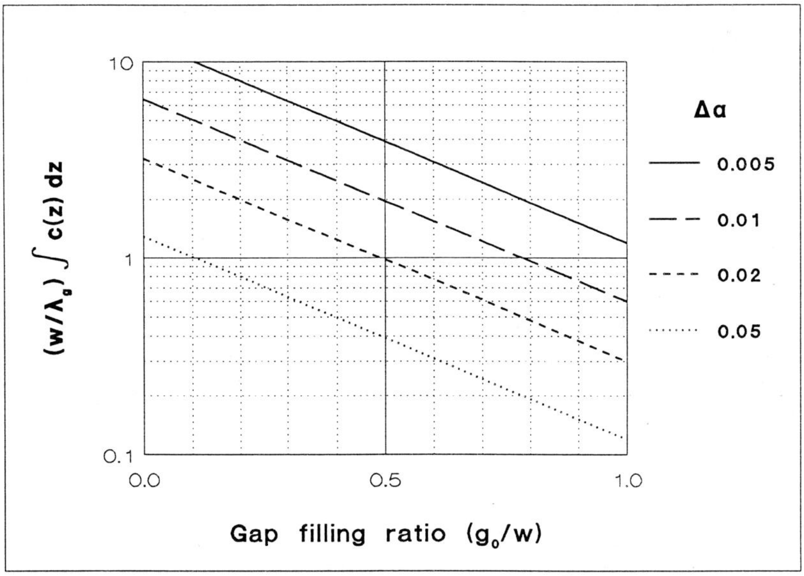

Figure 6.13 Accumulated coupling effect $\int cdz$ as a function of the filling ratio g_o/w for different inter-waveguide angles $\Delta\alpha$.

distance between two waveguides in the aperture plane amounts to *1/5* of the Gaussian width. At the center of the array the sum term will then amount to $2A_i\, e^{-(1/5)^2} \simeq 2A_i(1\text{-}0.04)$, in which the latter term will cause distortion because it will vary across the aperture.

From this example it is clear that coupling effects in the array will be smaller by at least one order in comparison to the value read from figure 6.13. The relative power in the distortion being proportional to $(\Delta A_i/A_i)^2$ it is seen that cross talk through coupling effects in the array will be considerably below -20 dB for a wide ranges of parameter values (in this example even less than $10\log 0.02^2 \simeq -34$ dB).

E. Phase transfer incoherence

The length of the array waveguides usually incorporates thousands of wavelengths. Small local deviations of the propagation constant may, therefore, lead to considerable phase transfer errors. If these errors are randomly distributed across the array waveguides, they will add a random component to the radiation pattern, and thus cause cross talk.

The tolerance requirements for the phase transfer may be estimated as follows. If the random phase error has a spread of σ_ϕ radians, and $\sigma_\phi \ll 1$, then the relative power in the random component follows as σ_ϕ^2. This power is randomly radiated into the effective numerical aperture θ_e of the waveguides, which amounts to $\lambda_o/(4w_e)$ in a Gaussian approximation (w_e is the effective modal width). In the image plane this power spreads over a width $W_f = 2\theta_e f = \lambda_f f/(2w_e)$, and a fraction w_e/W_f will be coupled into each receiver on average. We thus find as a coarse estimate of the cross talk level:

$$C = -10 \log_{10} (2\sigma_\phi^2 \frac{w_e^2}{f \lambda_f}). \qquad (30)$$

With $w_e/\lambda_f \simeq 5$ *and* $w_e/f \simeq 0.01$ as representative values, a spread $\sigma_\phi = 0.1$ in the phase transfer will yield a cross talk level in the order of -30 dB. A phase spread of 0.1 radian corresponds to a spread of $\lambda/60$ in the optical length of the array waveguides. With an internal wavelength of 0.5 µm, which is representative for InGaAsP waveguides operating at 1.55 µm, this corresponds to a deviation in the order of 10 nm for the optical channel length! This seems to be an unrealistic requirement for today's technology. It should be noted, however, that the requirement applies to the average optical length measured along the whole waveguide path, and that the different paths are located close to each other on the chip. The experiments reported in this chapter (section 6.7) demonstrate that coherence is preserved to a large extent through arrays with a length of several millimeters.

F. Mode conversion in the array waveguides

A final source of cross talk may result from mode conversion at imperfections in the waveguides or at the junctions between straight and curved waveguides. Because the latter discontinuities occur in pairs, power can be transferred to and back from unwanted modes (usually the first-order mode), thus creating a second path with a different optical length. Power carried by this path should be small, generally less than 1%, which means that power conversion per junction should maximally amount to a few percents.

6.4.4 Polarization dependence

In waveguide systems with polarization dispersion, the array performance will be polarization dependent. Without special precautions the directions for the *TE*- and the *TM*-polarized beam(s) will be different, as depicted in figure 6.14. Polarization independence will occur if the *TE*-pattern with order *m* coincides with the *TM*-pattern with order *m-i*. This will happen if $m\lambda_{TE}$ equals $(m-i)\lambda_{TM}$, or

$$m\,\Delta\lambda_{pol} = -\,i\,\lambda_{TM}\,, \qquad (31)$$

no minus sign

in which the polarization dispersion $\Delta\lambda_{pol}$ is defined as:

$$\Delta\lambda_{pol} = \lambda_{TM} - \lambda_{TE}\,. \qquad (32)$$

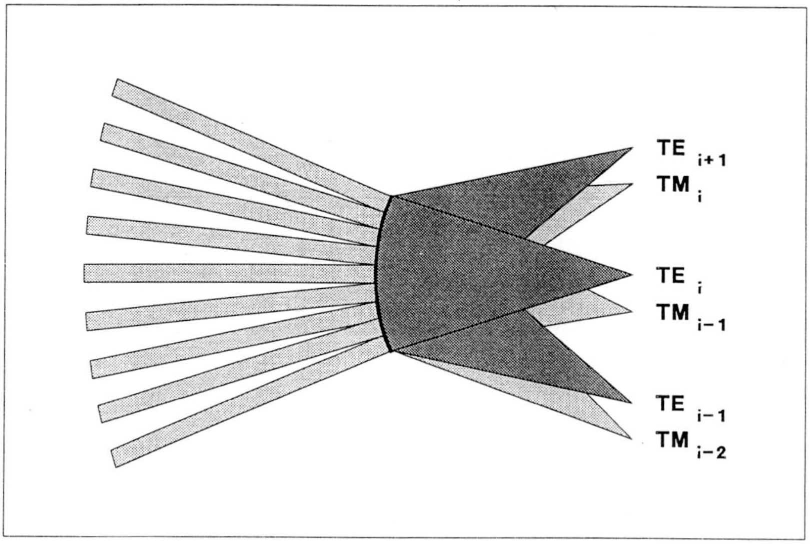

Figure 6.14 *Schematic picture of the different orders, both TE- and TM-polarized, which are radiating from the array.*

Equation 6.31 can be satisfied through a proper choice of array order m and polarization dispersion, as will be shown in the next section.

Since the orders for *TE*- and *TM*-polarization differ, it follows from equation 6.15a that the dispersion for each polarization will be slightly different, so that total polarization independence is not possible. If $m \gg i$, which will normally be the case, the dispersion difference will be small and virtual polarization independence may be obtained.

6.5 Design

6.5.1 Design requirements

If the absolute phase transfer of the component is not relevant, which will normally be the case, a demultiplexer or polarization splitter is fully characterized from an application point of view by the following quantities:

- the **number** n_c **of output channels**
- the **channel wavelengths** λ_i (assumed to be spaced equidistant)
- the **insertion loss** L_i of the i-th receiver channels
- the **cross-talk level** C_{ij} in the i-th channel (relative to the input level) when channel j is excited.

For a wavelength demultiplexer the latter three quantities will differ for TE- and TM-polarized light. The polarization-dependence of the insertion loss and the maximum cross talk have to be minimized if polarization-independence is required.

If the circuit has to be produced in large quantities the device size is also of interest. In the following sections the design of the phased array, so as to meet requirements with respect to the properties described above will be discussed.

6.5.2 Waveguide parameters

The device size is, to a large extent, determined by the choice of the waveguide, notably the waveguide width and the effective index contrast. High effective index contrasts allow for short bending radii to be employed, thus reducing component size.

For a given maximal insertion loss for the outer channels, the length of the free propagation region scales with the distance between the outer receiver waveguides in the image plane. This distance can be reduced both by reducing the waveguide width and by increasing the (lateral) index contrast.

To minimize device size the waveguide width should, therefore, be chosen to be as small as possible, and the index contrast as large as possible. With waveguide widths in the order of 1 μm and bending radii as short as 100 μm, a device size far into the submillimeter range is feasible.

6.5.3 Receiver spacing

Once the waveguide parameters have been chosen the receiver spacing d_r can be determined from the required channel isolation using figures 6.10 and 6.11. Some margin should be included to allow for contributions from other cross-talk causing mechanisms, as discussed in section 6.4.3. Normally, receiver spacings (on centerlines) in the order of two or three times the waveguide width will prove sufficient.

6.5.4 Free propagation length

The minimal free propagation length f_{min} is bounded by the maximally allowed insertion loss of the outer channels; it is found from:

$$f_{min} = \frac{y_{max}}{\theta_{max}}, \tag{33}$$

in which y_{max} is the distance of the outer receiver to the center of the image plane and the maximal dispersion angle θ_{max} is found from the maximally acceptable angular insertion loss L_{max} by inverting equation 6.19:

$$\theta_{max} \simeq \theta_o \sqrt{L_{max} / 8.7}, \tag{34a}$$

in which θ_o is the equivalent Gaussian width of the single-waveguide far field. Using equation 6.17 this equation can be rewritten as

$$\theta_{max} \simeq 0.135 \frac{\lambda_f}{w_e} \sqrt{L_{max}}, \tag{34b}$$

in which w_e is the effective width of the modal field and λ_f is the wavelength within the free propagation region.

It should be noted that in the above formula L_{max} is the maximal angular diffraction loss, as described by formula 6.19. The central channel loss L_o (equation 6.22), which is determined by the ratio w_e/d, adds up to this loss, and should first be subtracted from the maximal acceptable loss in order to arrive at the maximum angular loss L_{max}.

6.5.5 Array parameters

Once the receiver spacing d_r in the image plane has been determined, the relative dispersion δy of the array (equation 6.15b) should have the following value:

$$\delta y = \frac{d_r}{\delta\lambda}, \tag{35}$$

in which $\delta\lambda$ is the relative channel spacing $\Delta\lambda_g/\lambda_c$ (equation 6.15c). From equation 6.15b it can be seen that *(if $m \simeq 1$)* the relative dispersion equals the first derivative $dl/d\alpha$ of the element length (note that $m\lambda_c$ is the difference in length between adjacent elements), so that the element length can be written as

$$l(\alpha) = l_o + \alpha\delta y, \tag{36}$$

in which the constant l_o can still be chosen freely. The corresponding straight section length s and the radius R of the curved section follow from equations 6 .8 and 6.9. The values of l_o and the receiver-to-transmitter distance L (equation 6.11) are fixed if we select a value s_r and R_r for s and R at an (arbitrary) reference angle α_r:

$$l_o = l_r - \alpha_r \delta y = 2s_r + 2\alpha_r R_r - \alpha_r \delta y, \tag{37}$$

$$L = 2s_r \cos \alpha + 2R_r \sin \alpha. \tag{38}$$

After L and l_o are fixed the last parameter to be determined is the center angle α_c of the array (which can be chosen independent of α_r). The design is thus seen to have three degrees of freedom: s_r, R_r and α_c. This freedom can be used to optimize the design.

To arrive at a suitable design it is convenient to make a graph of $s(\alpha)$, $R(\alpha)$, and the element height $h(\alpha)$:

$$h(\alpha) = s(\alpha) \sin \alpha + R(\alpha)\,(1 - \cos \alpha), \tag{39}$$

over the range from 0° to 180°, for one or more combinations of s_r and R_r. Figure 6.15 shows an example. To be realizable the following constraints must be satisfied over a sufficiently large angular interval:

$$s > f_{min}, \tag{40a}$$

$$R > R_{min}, \tag{40b}$$

$$\frac{dh}{d\alpha}\Delta\alpha > w * factor \quad \text{with } factor \text{ between 2 and 3}, \tag{40c}$$

in which f_{min} is the minimal value for the length of the free propagation region, as described in equation 6.33, and R_{min} is the minimal radius for which the bending losses are still acceptable. The last constraint (6.40c) avoids elements coming to close to each other.

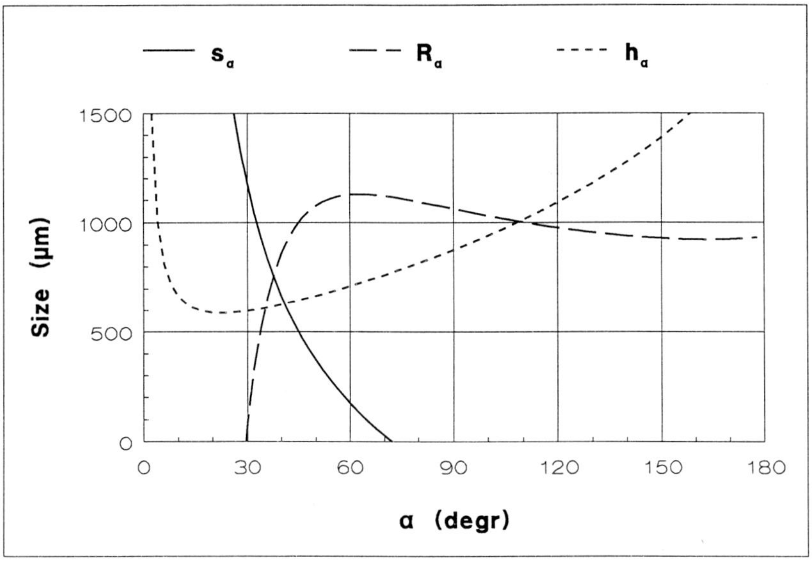

Figure 6.15 The phased-array straight section length s, radius R and element height h for s_r = 500 μm and R_r = 1000 μm at α_r=45°.

From the figure it can be seen that there is only a small region (between 30° and 70°) where all constraints can be satisfied simultaneously. To enlarge this region other values of s_r and R_r have to be tried. A strategy for arriving at an optimal choice of these parameters will be explained in the next section. In this section we will proceed with the determination of the other array parameters from a graph similar to figure 6.15 once the values for s_r and R_r have been selected.

The angular array aperture $W_\alpha = \alpha_n - \alpha_1$ (see figure 6.6b, n is the total number of array elements) should be chosen so as to keep the spillover, and the related cross talk, sufficiently small. If it is chosen 4 times the equivalent Gaussian width of the far field, W_α follows as:

$$W_\alpha = 4\theta_o = \frac{4}{\sqrt{2\pi}} \frac{\lambda_f}{w_e} \simeq 1.6 \frac{\lambda_f}{w_e}. \quad (41)$$

The central array angle α_c has to be chosen such that the constraints 6.40a-c are satisfied over the interval $\alpha_c \pm ½W_\alpha$. If this is not possible, reduction of W_α or another choice of s_r and R_r must be considered.

If the required array aperture W_α is smaller than the available interval for which $s>f_{min}$ we can choose f greater than f_{min}, thus further reducing the outer-channel insertion loss at the expense of an increased number of array waveguides. In designing a wavelength demultiplexer this freedom can also be used to obtain polarization-independent operation, as will be discussed in the next section.

The order m of the array is found (from equations 6.12 and 6.15) as

$$m = \frac{d\,\delta y}{r f \lambda_c}, \quad (42)$$

by rounding the right-hand side to the nearest integer value[1]. The angular element spacing $\Delta\alpha$ then follows as:

$$\Delta\alpha = mr\lambda_c/\delta y. \quad (43)$$

---

1) The effect of rounding the right-hand side of equation 6.42 leads to a slight adaption of the free propagation length f through the choice of $\Delta\alpha$ according to equation 6.43.

The corresponding number of array waveguides is

$$n_a = \frac{W_\alpha}{\Delta\alpha} + 1, \tag{44}$$

which completes the array design. The component is completed by adding the transmitter and receiver waveguides. Figure 6.16 shows two different implementations. For optimal coupling efficiency the receiver waveguides should be located as indicated in figure 6.17.

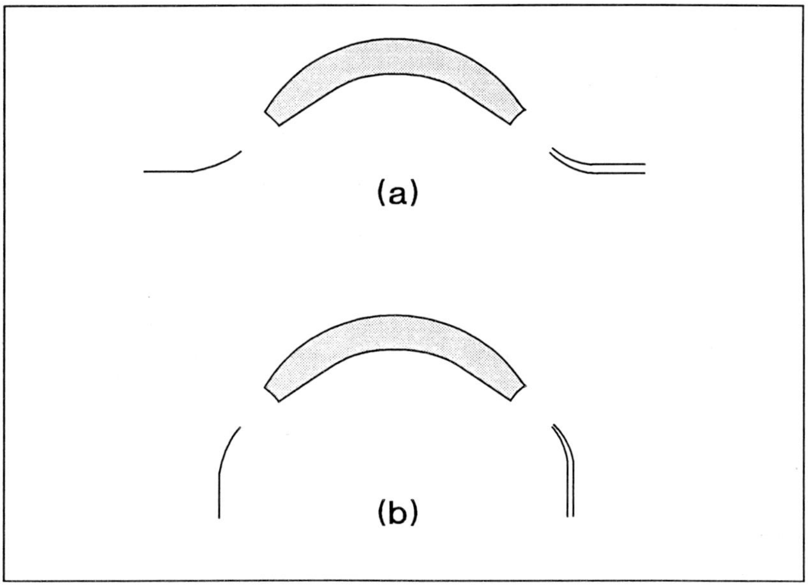

Figure 6.16 *Two different implementations of a phased array:*
a) *with in-line input and output waveguides,*
b) *with input and output waveguides at the same side of the chip.*

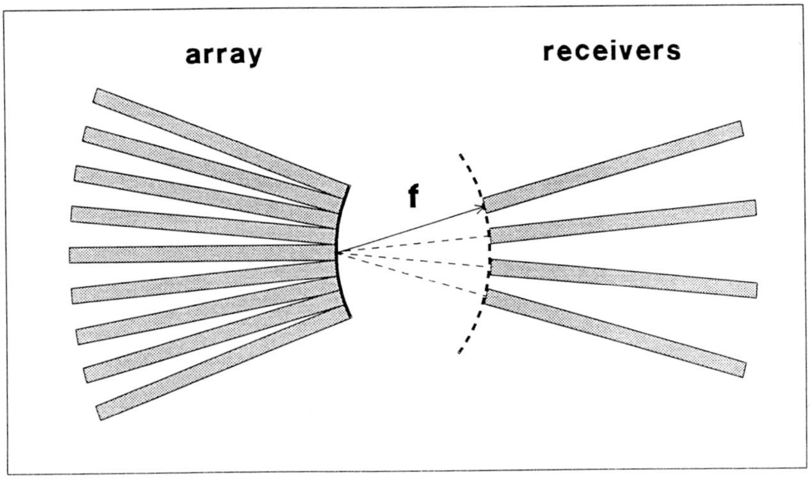

Figure 6.17 Optimal location of the the receiver waveguides.

6.5.6 Design optimization

From the curves of figure 6.15 the bending radius R is seen to vary considerably as a function of α. For small values of R it will be necessary to account for the change in β due to waveguide bending. It is advantageous to keep the variation of R as small as possible in order to reduce the dependence of the array transfer on bending effects.

Two of the three degrees of freedom (s_r, R_r, and α_c) can be used for flattening the curve $R(\alpha)$ by requiring $dR/d\alpha$ and $dR^2/d\alpha^2$ to be zero at the reference angle α_r (which can be freely chosen). Upon calculation we find the following expressions for these derivatives:

$$\frac{dR}{d\alpha} = \left(s - \frac{\tfrac{1}{2}\delta y}{\tan\alpha}\right) / \left(1 - \frac{\alpha}{\tan\alpha}\right), \qquad (45)$$

$$\frac{d^2 R}{d\alpha^2} = \tfrac{1}{2}\delta y \left(1 + \frac{1}{\sin^2 \alpha}\right) - R. \tag{46}$$

In the latter formula it has been assumed that $dR/d\alpha=0$. From these formulae it is seen that the first derivative can be put to zero by choosing

$$s_r = \frac{\tfrac{1}{2}\delta y}{\tan \alpha_r}, \tag{47}$$

and the second one by choosing

$$R_r = \tfrac{1}{2}\delta y \left(1 + \frac{1}{\sin^2 \alpha_r}\right). \tag{48a}$$

Empirical analysis reveals that $R(\alpha)$ is reasonably flat over a wide range of R_r-values, and that an even wider flat range is obtained with

$$R_r = \frac{\tfrac{1}{2}\delta y}{\sin^2 \alpha_r}, \tag{48b}$$

at the cost of a slight decrease in flatness. Inspection of formulae 6.8, 6.9, and 6.36 to 6.38 reveals that all array dimensions depend linearly on δy after substitution of formulae 6.47 and 6.48a or 6.48b. This means that all array dimensions can be normalized on the relative dispersion δy, which considerably simplifies analysis. Figure 6.18 shows the normalized section length, bending radius, element height and its derivative of an array designed for a flat R-curve at $\alpha_r=45°$. To arrive at the actual dimensions all values should be multiplied by δy.

From the figure it can be seen that s decreases approximately linearly for angles beyond the reference value $\alpha_r=45°$, and is positive from approximately 30° to 90°. The height derivative $dh/d\alpha$, which is indicative of the distance between the array elements, has a minimum slightly beyond the reference value α_r. The general shape of the curves remains the same over a wide range of α_r-values.

Figure 6.19 shows the width W_+ of the angular interval over which s is positive, the straight section length s_r, the bending radius R_r, and the minimum of the height derivative $dh/d\alpha$, as a function of the reference radius. With this graph the array can be dimensioned if we know the minimal free propagation length f_{min}.

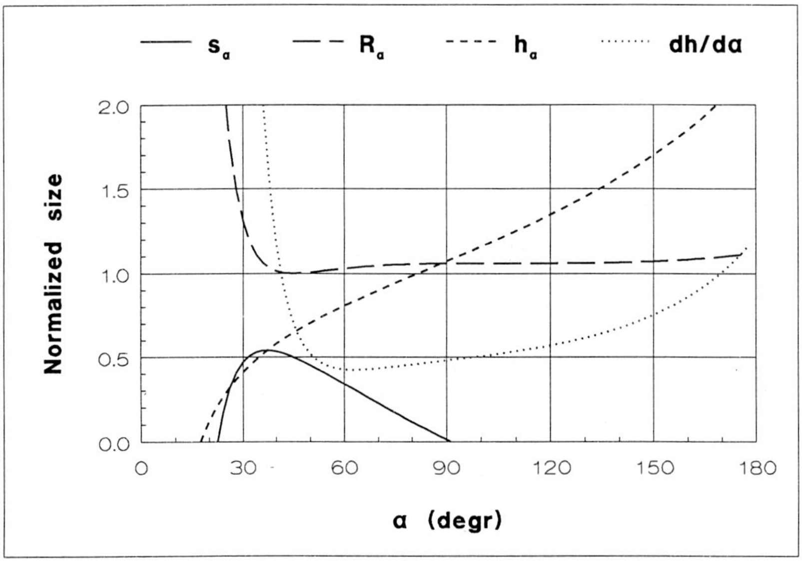

Figure 6.18 The section length s, radius R, element height h, and the height derivative dh/dα, normalized on the relative dispersion δy, with R(α) flattened around α=45°.

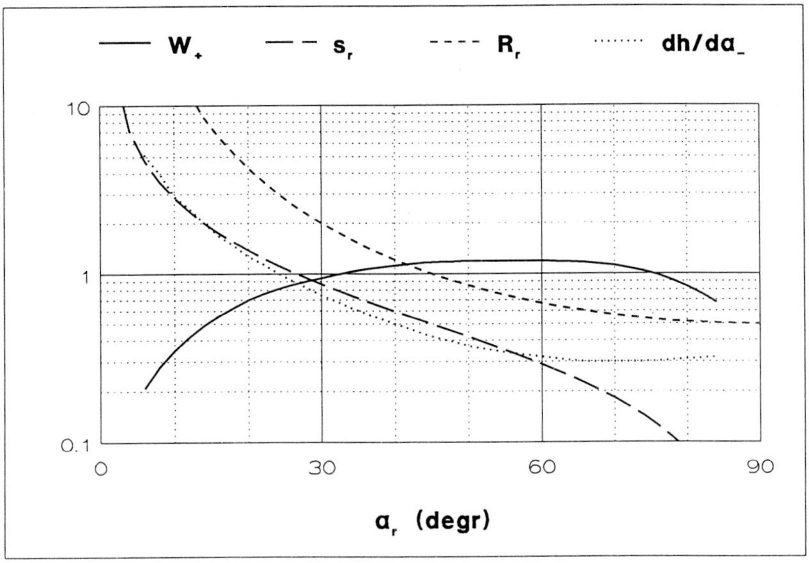

Figure 6.19 Design curves for the selection of the reference angle α_r: (curve 1: radians, curves 2-4 normalized on δy)

> *1) width W_+ of the angular interval with positive s-values,*
> *2) straight section length s_r for flattening $R(\alpha)$,*
> *3) bending radius R_r for flattening $R(\alpha)$,*
> *4) minimum of the height derivative $dh/d\alpha$.*

Figure 6.20 shows the dependence of the normalized array length and the height of the reference element, which are indicative of the array dimensions, on the choice of the reference angle α_r. The contributions of the transceiver waveguides to the array dimensions are readily computed.

If the approach described above does not yield a satisfactory design, the requirement for a flat R(α) curve can be abandoned, or another array geometry can be selected. The interdependence of the various parameters being rather complicated, another design strategy will have to be developed in order to arrive at optimal designs.

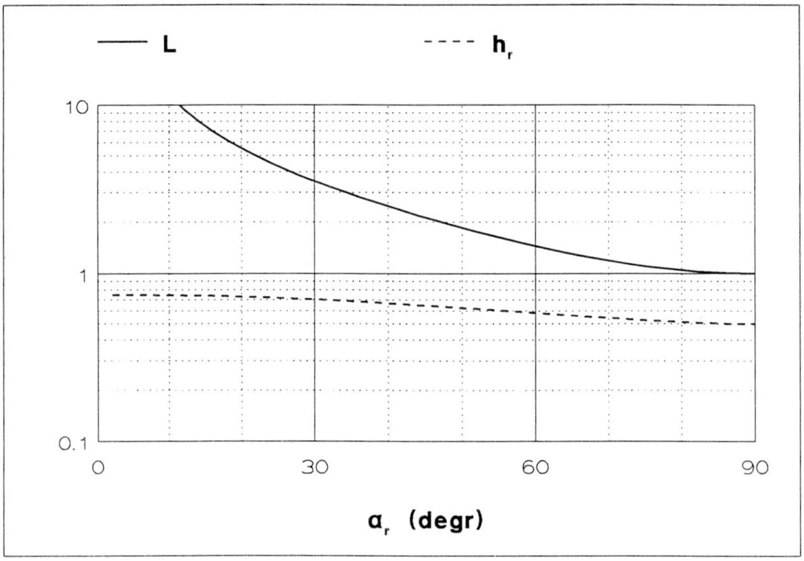

Figure 6.20 Normalized length of the array (from transmitter to receiver), and the height of the reference element, as a function of the reference angle.

6.5.7 Correction for bending effects

According to the present approach the phase transfer of the curved waveguides is described as $\phi = 2\alpha R\beta_g$, in which β_g is the straight-waveguide propagation constant. The actual phase transfer will equal $2\alpha\beta_\phi(R)$ in which β_ϕ is the angular propagation constant as discussed in chapter 4.

If the applied bending radii are in a range in which the shift of the mode-profile towards the outer edge of the bend can no longer be neglected, the actual phase transfer will differ from its design value and thus degrade array performance. A simple solution to this problem is to apply a correction to the radius R, such that:

$$2\alpha\beta_{\phi}(R') = \Phi = 2\alpha R\beta_{g}, \tag{49}$$

in which R' is the corrected value. It is found as follows. If we neglect bending effects, the actual phase transfer $2\alpha\beta_{\phi}(R)$ seems to be caused by a waveguide with radius $R''=\beta_{\phi}(R)/\beta_{g}$, which will be greater than R by an amount ΔR_{ph}. The corresponding phase error can be neutralized by reducing R by the same amount:

$$R' = R - \Delta R_{ph}, \tag{50a}$$

in which

$$\Delta R_{ph} = \frac{\beta_{\phi}(R)}{\beta_{g}} - R. \tag{50b}$$

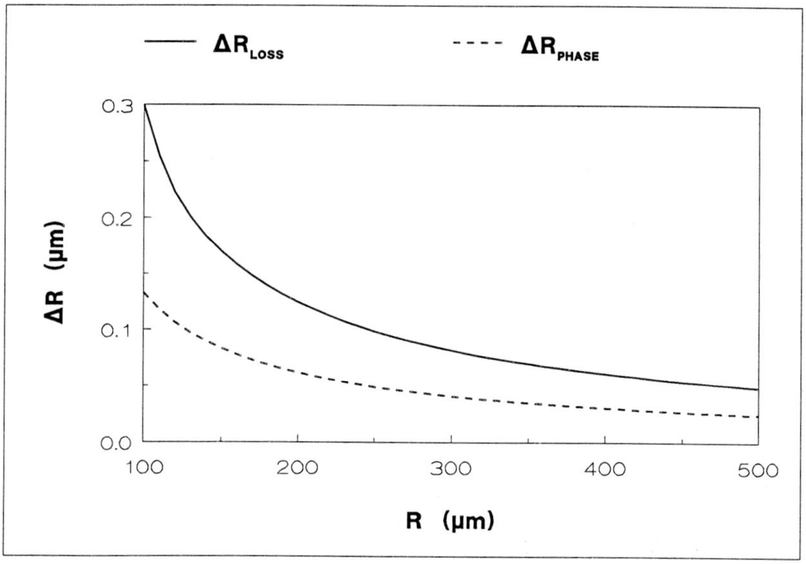

Figure 6.21 Optimal offsets (between centerlines) for minimal coupling loss (a) and minimal phase transfer error (b) as computed for the example presented in section 6.5.9.

The computed offset ΔR_{ph} differs from the optimal offset for minimal coupling loss ΔR, as described in chapter 4. Figure 6.21 shows both offsets (between centerlines) for the example described in section 6.5.9. From the figure it can be seen that the offset required for constant phase transfer is approximately equal to half the offset required for minimal coupling loss. As long as the latter offset is in the order of a few tenths of a micron, the loss penalty in applying the constant-phase-transfer offset of equation 6.50b will be negligible. If the variation of R is small, a single correction can be applied to all radii.

If necessary, coupling loss may be further reduced by optimizing both the offset and the curved-waveguide width. If very small bending radii are to be employed, bending effects will have to be accounted for in choosing the array parameters. This complicates design. In the present approach this problem is circumvented: the design parameters are computed as if there were no bending effects and corrections are applied afterwards to the bending radius, while leaving all other parameters unchanged.

6.5.8 Polarization-independent demultiplexer design

In section 6.4.4 it was shown that polarization independence can be achieved through a proper choice of the array order m. From equation 6.15b it is seen that for a given dispersion δy the array order m is exclusively determined by the free propagation length f through the relation $\Delta\alpha = d/f$. Combination of these relations with equation 6.31 gives us the following criterion for polarization independence:

$$f = \frac{d}{i}\frac{\Delta\lambda_{pol}}{\lambda_{TM}}\frac{dx}{d\lambda} \simeq \frac{d}{i}\frac{\Delta\lambda_{pol}}{\lambda_c^2}\delta y, \tag{51}$$

in which i is an arbitrary integer. If the range of possible values for f, as discussed in section 6.5.5, contains a value satisfying the above relation, polarization independent design is feasible.

6.5.9 Design example

As an example we will discuss the design of a polarization splitter for operation at *1.5500* μm, to be realized in the aluminum oxide ridge-guide system described in chapters 2 and 3.

The design will be developed for the smallest waveguide width which can be fabricated reproducibly in our laboratory: *2 μm*, and an etch depth of *350 nm*, which allows for low-loss bending radii as short as *150 μm*. Figure 6.22 shows the waveguide geometry.

An effective index of *1.5362* has been found for the TE_{00} mode and *1.5180* for the TM_{00} mode using the effective index method. These correspond to wavelengths $\lambda_{TE} = 0.9894$ and $\lambda_{TM} = 1.0013$ within the waveguide. The central wavelength is thus $\lambda_c = 0.9953$ and the relative wavelength spacing

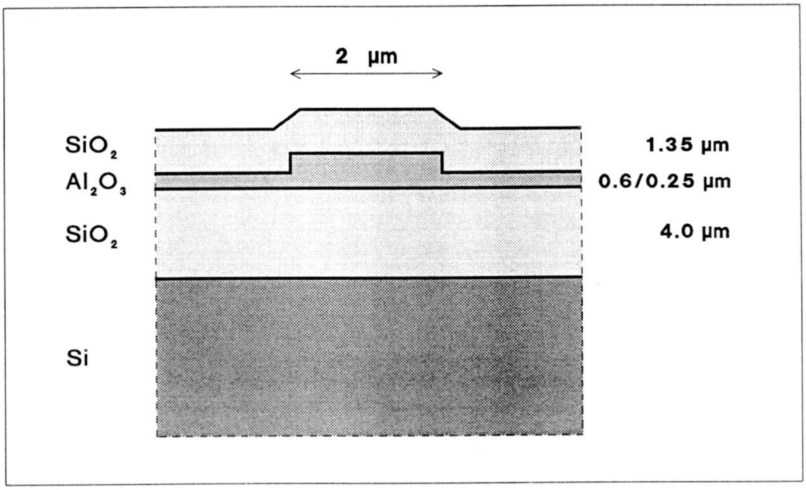

Figure 6.22 Waveguide geometry used in the example.

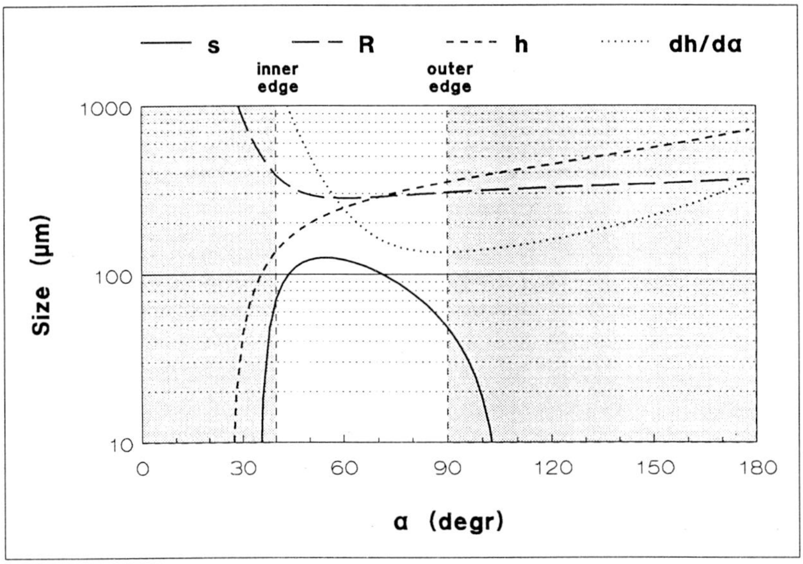

Figure 6.23 The straight section length s, radius R, element height h and height derivative dh/dα for δy=480 μm and α_r=60°.

$\delta\lambda$ = *1.2%*. The lateral *V*-parameter equals *3.8* for *TE*-polarized and *3.6* for *TM*-polarized light, so that the waveguides are bimodal.

From figure 6.10 it can be seen that a ratio $g/w=1.5$ provides more than *40 dB* isolation between the output channels, so that the channel spacing on centerlines amounts to *5 μm*. This requires a relative dispersion $\delta y = (g+w)/\delta\lambda = 5/0.012 = 420\ \mu m$.

From formula 3.11 we find an effective modal width of *1.63 μm*. If the resolution of the lithographic process is *1 μm* the central channel insertion loss L_o (equation 6.22) is *0.5 dB*. For a maximum total insertion loss of *1 dB* the margin for the angular dispersion loss is *0.5 dB*. The corresponding maximum dispersion angle θ_{max} (formula 6.34) is *0.15 radians*, from which f_{min} (equation 6.33) follows as $3/0.15 \simeq 17\ \mu m$. The angular array aperture W_α corresponding to $4\theta_o$ (equation 6.41) amounts to *0.9 radians (50°)*.

Based on these values a reference angle of *60°* is selected from figure 6.19. Figure 6.23 shows the array parameters for $\delta y=420\ \mu m$ and $\alpha_r=60°$. If we choose the array aperture from *40° to 90°* (the non-shaded region in figure 6.23, $\alpha_c = 65°$), the free propagation length f can be chosen to be *40 μm*, leading to an even lower angular insertion loss of *0.1 dB*, so that the totally expected diffraction loss amounts to *0.6 dB*. The array order found from equation 6.42 is *30*, the required angular waveguide spacing $\Delta\alpha$ (equation 6.43) amounts to *0.075* radians *(4.3°)*.

The corresponding number of waveguides n_a (equation 6.44) is *13*. From figure 6.13 it is seen that no coupling problems are to be expected in the array with the value of $\Delta\alpha$ found above ($\int c(z)dz < 0.2$). From figure 6.23 the radius is seen to vary between *280 μm* in the center and approximately *400 μm* at the inner edge (the left side of the non-shaded window). Figure 6.21 shows that for this range an offset of *0.07 μm* will combine the required phase transfer with low coupling loss.

The array length L is read from figure 6.20 as *600 μm*. If the transmitter and the receiver are realized using R=*300 μm* bends, the total length will be *1150 μm* in the implementation of figure 6.16a. The maximum array height is found from figure 6.20 to be *400 μm*. Figure 6.24 shows the mask-layout. Smaller dimensions can be obtained by increasing the effective index contrast, which will reduce both the receiver spacing d_r and the minimum bending radius.

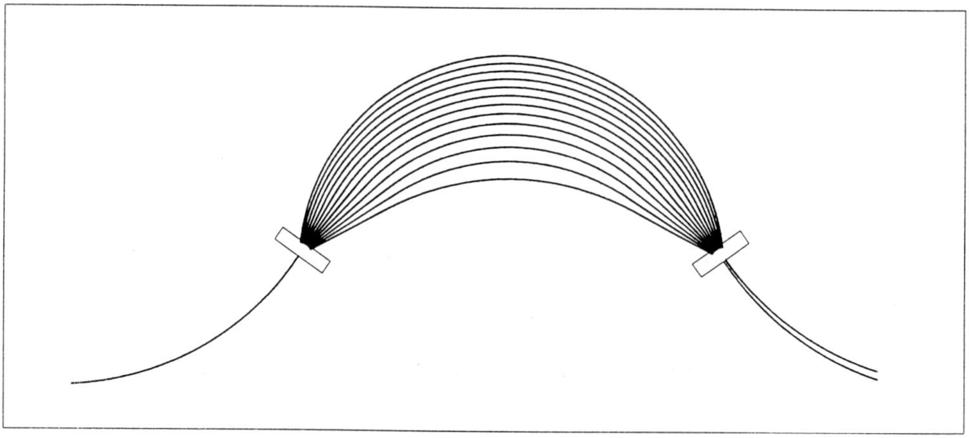

Figure 6.24 *Mask layout of the polarization splitter described in the example.*

6.6 Simulation

6.6.1 Network description

From a circuits point of view a demultiplexer or a polarization splitter can be conceived of as an *(m+1)*-port (see figure 6.25a), in which *m* is the number of output channels (equaling 2 for a polarization splitter). If reflections are neglected, the component can be described by a transmission matrix *T*:

$$U_r = T\,U_t, \tag{52a}$$

$$U_r = \begin{pmatrix} u_{r1} \\ .. \\ u_{rm} \end{pmatrix}, \qquad T = \begin{pmatrix} t_1 \\ .. \\ t_m \end{pmatrix}, \qquad U_t = u_t, \tag{52b}$$

in which the complex signal amplitudes u_{ri} in the *i-th* receiver waveguide and u_t in the transmitter waveguide are represented as:

$$u_s = \sqrt{P_s}\, e^{j\phi_s}, \tag{53}$$

Normalizing the signal amplitude u_s with respect to the root of the signal power P_s has the advantage of making it invariant to the specific structure of the waveguide. This is advantageous if waveguides with different geometries (widths) are used.

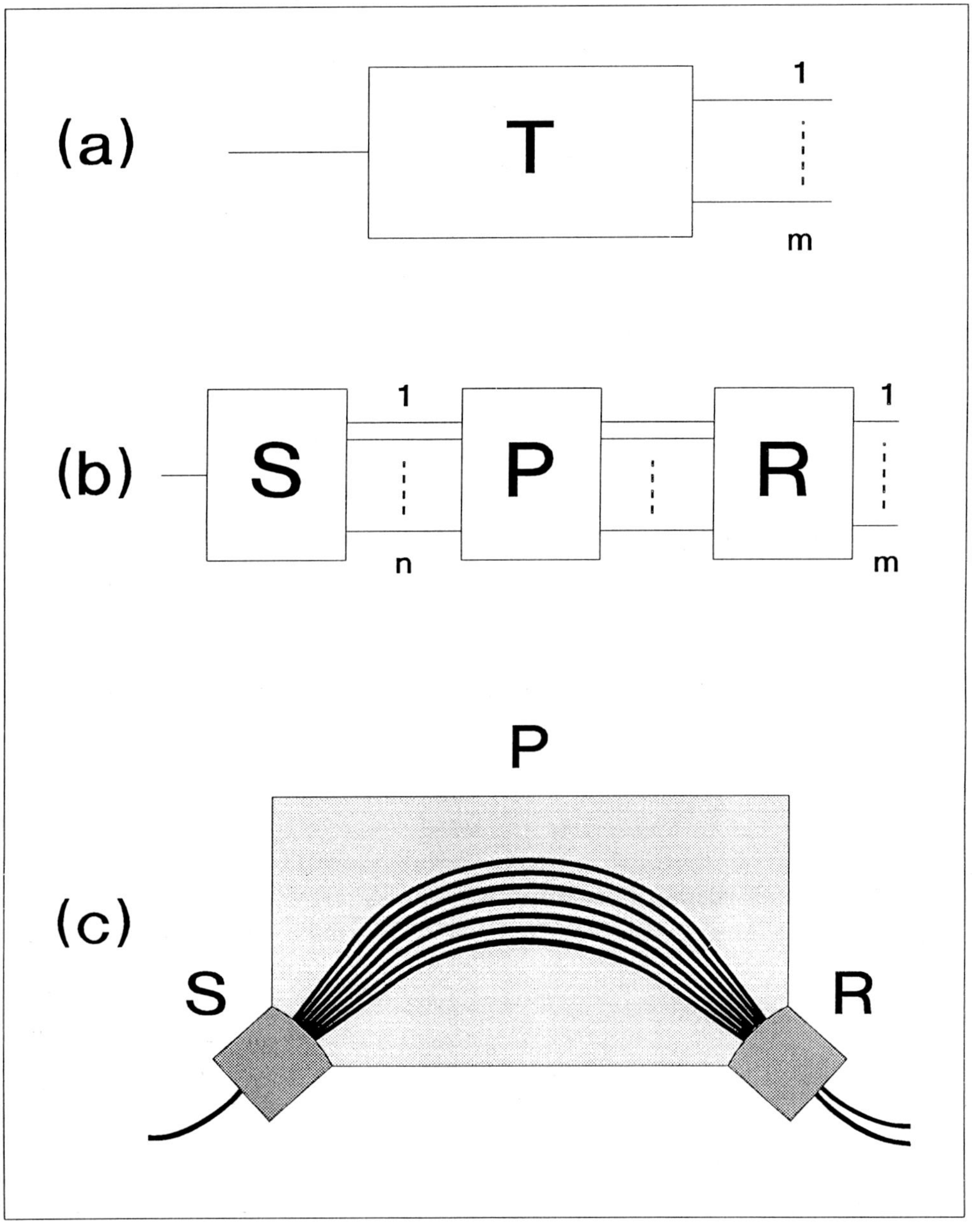

Figure 6.25 Network representation of a phased array.

All elements t_i of the transmission matrix are functions of both wavelength and polarization. A general description, which covers both the application as a wavelength demultiplexer and as a polarization splitter, can be obtained by describing all matrix elements as a function of the wavelength λ_g within the waveguide, or its reciprocal, the propagation constant β_g. For the response of the component it does not matter whether a change in λ_g or β_g is caused by a frequency change or by a change of the polarization.

The overall transmission matrix T can be decomposed as follows (figure 6.25b):

$$T = R\,P\,S\,, \tag{54a}$$

with

$$R = \begin{pmatrix} r_{11} & .. & r_{1n} \\ .. & .. & .. \\ r_{m1} & .. & r_{mn} \end{pmatrix}, \quad P = \begin{pmatrix} p_{11} & 0 & 0 \\ 0 & .. & 0 \\ 0 & 0 & p_{nn} \end{pmatrix}, \quad S = \begin{pmatrix} s_1 \\ .. \\ s_n \end{pmatrix}, \tag{54b}$$

in which s_i describes the transmission from the transmitter to the *i-th* array channel, p_{ii} describes the propagation through the *i-th* array waveguide, and r_{ij} describes the transmission form the *i-th* array channel to the *j-th* receiver channel. The representation of the propagation matrix P as a diagonal matrix assumes that there is no coupling between the array waveguides.

In this description the insertion loss of the *i-th* receiver channel follows as

$$L_i = -20\,\log_{10} | t_i(\beta_i) | \,, \tag{55}$$

and the channel cross talk C_{ij} , which is defined as the the *channel-j* signal level in receiver i, as

$$C_{ij} = 20 \log_{10} | t_i(\beta_j)|. \tag{56}$$

In the following paragraphs the computation of the matrices ***T***, ***P***, and ***R*** will be described.

6.6.2 Transmitter-to-array transfer

The transmitter-to-array transfer comprises four phenomena: transmission and reflection at the end of the waveguide, diffraction through the free propagation region, coupling into the array and propagation through the first part of the array. In this part of the array coupling effects can not be neglected. To facilitate computations the effects of coupling at both ends of the array will be included in the matrices ***S*** and ***R*** describing the transmitter and receiver *sub-circuit*, so that the array itself can be considered to be uncoupled.

The source field $U_1^+(y)$ in the transmitter aperture plane S_1^+ (see figure 6.26) is found as:

$$U_1^+(y) = \tau\, U_1^-(y), \tag{57}$$

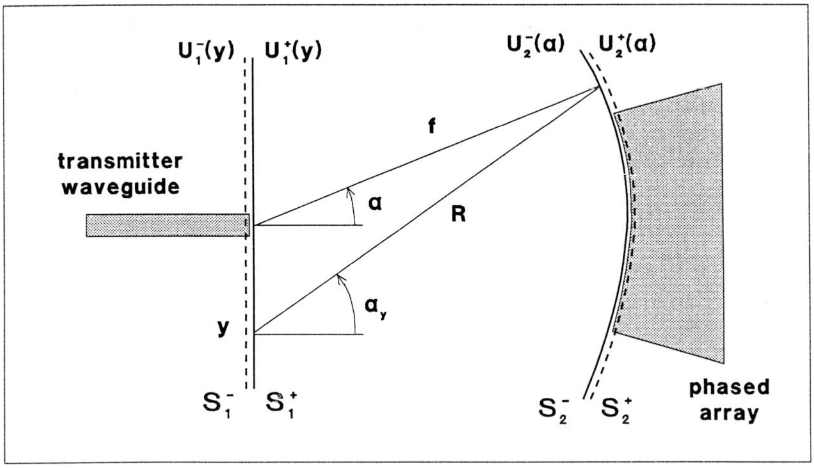

Figure 6.26 Local coordinate system at the transmitter, centered around the array central angle α_c.

in which $U_1^-(y)$ is the lateral mode profile of the transmitter waveguide, and τ is the transmission coefficient through the transmitter aperture plane. It may be estimated using the Fresnel formula applied to the effective indices of the mode on one side, and the (two-dimensional) plane wave on the other side of the junction. For low to moderate index contrasts the transmission coefficient will be close to unity. It will be omitted in the following formulas.

The diffraction field $U(\alpha,\rho)$ is computed with the Rayleigh-Sommerfeld approximation:

$$U_2^-(\alpha) = U(\alpha,f) \simeq \tfrac{1}{2}\beta_f \int_{S_1^+} H_o^{(2)}(\beta_f R)\, U_1^+(y)\cos(\alpha_y-\alpha_c)\, dy, \qquad (58a)$$

in which β_f is the plane-wave propagation constant in the film between the transmitter and the phase array (the free propagation region). The argument $(\alpha_y - \alpha_c)$ of the cosine function is the angle between the direction from the source point y to the observation point (α, f) and the transmitter axis, which coincides with the array central angle α_c. It is slightly dependent on y but approximately equal to $(\alpha - \alpha_c)$. Figure 6.26 shows the local coordinate system centered around α_c. The Green function $H_o^{(2)}(\beta_f R)$ can be approximated as:

$$H_o^{(2)}(\beta_f R) \simeq \frac{(1+j)}{\sqrt{\pi \beta_f R}} e^{-j\beta_f R}, \qquad \text{for } \beta_f R \gg 1. \tag{58b}$$

In order to find the excitation coefficients u_i of the individual array waveguides, we assume that these coefficients are proportional to the amplitude of the diffraction-field at the center of the waveguide:

$$u_i = a\, U_2^-(\alpha). \tag{59a}$$

The proportionality constant a follows from the requirement that:

$$\Sigma u_i^2 = \eta P_2^-, \tag{59b}$$

in which η is the coupling efficiency from the diffraction field to the guided modes in the array, and P_2^- is the power in the diffraction field at S_2^-, which is related to the transmitted power $P_t\,(=P_1^-)$ through:

$$P_2^- = e^{-\alpha_f f} P_t . \qquad (60)$$

The exponential accounts for the propagation loss in the free propagation region; it will be omitted in the following formulae. The proportionality constant a is found (from equations 6.59 and 6.60) to be:

$$a \simeq \frac{\eta P_t}{\sum_i U^2(\alpha_i, f)} . \qquad (61)$$

For the computation of the coupling efficiency η we assume that the guided field in the plane S_2^+, at the beginning of the array, may be approximated as the sum of the individual waveguide modes U_m:

$$U_2^+(\alpha) = C \sum_i U_m(\alpha - \alpha_i) . \qquad (62)$$

in which α_i is the angular coordinate of the *i-th* waveguide. The field is illustrated in figure 6.27. The proportionality constant C is not relevant for the determination of the coupling efficiency because it occurs in the numerator as well as in the denominator of the overlap integral formula:

$$\eta = \frac{\left| \int U_2^-(\alpha)\, U_2^+(\alpha)\, d\alpha \right|^2}{\int \left| U_2^-(\alpha) \right|^2 d\alpha \int \left| U_2^+(\alpha) \right|^2 d\alpha} . \qquad (63)$$

Due to the overlap of the individual waveguide modes in the plane S_2^+, the field U_2^+ will not simply equal the sum of the individual modes. When propagating through the fan-out section, the guided power will be increasingly concentrated within the waveguide so that the field strength at the waveguide axis will increase. Consequently it will decay in the reverse direction. The field strength in the array aperture will thus be smaller than expected from simple summation of the uncoupled waveguide modes.

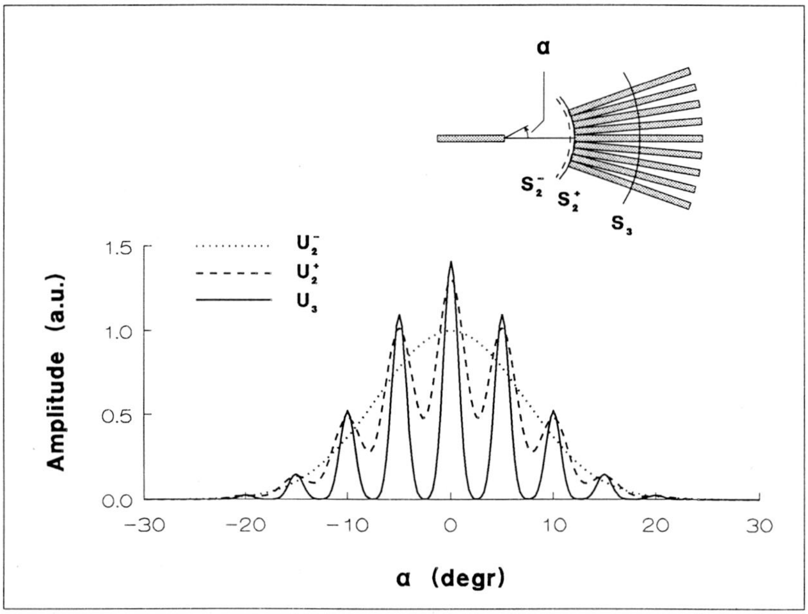

Figure 6.27 The field near the aperture of the phased array:

U_2^-: *Diffraction field at the array aperture*

U_2^+: *Sum field of the individual modes in the aperture*

U_3: *Sum field of the individual modes in the uncoupled part of the array.*

The need to analyze this effect is circumvented by applying the approach outlined above. The matrix elements s_i are thus found to be:

$$s_i = a\,U(\alpha_i, f), \tag{64}$$

which completes the computation of the transmitter array S.

6.6.3 Array transfer

If the effect of waveguide coupling on the excitation coefficients is included in the matrices describing the transmitter and receiver *sub-circuits*, the propagation matrix P of the phased array is a *nxn* diagonal matrix with elements

$$p_{ii} = e^{-j\beta_g l_i} \tag{65a}$$

$$p_{ij} = 0 \qquad i \neq j \tag{65b}$$

If mutual coupling effects in the array are small, but not negligible, their effects on the array transfer may be simulated by multiplying the propagation matrix on both sides with a coupling matrix:

$$P_c = C^T P\,C \tag{66}$$

in which T denotes transposition. If coupling effects are small, the elements of the coupling matrix follow as (see section 6.4.3D)

$$c_{ij} = 1 \qquad |i-j| = 0 \tag{67a}$$

$$= -j \int c(z)dz \qquad |i-j| = 1 \tag{67b}$$

$$= 0 \qquad |i-j| > 1 \tag{67c}$$

in which the coupling integral is computed from equation 6.29.

6.6.4 Array-to-receiver transfer

A straightforward way to analyze the receiver side of the component is by reconstructing the field in the array output aperture as the sum of the individual modes (corrected for coupling effects), computing the image field using the Rayleigh-Sommerfeld integral, and overlapping this field with the modal fields of the receiver waveguides in order to find the excitation coefficients of the receiver waveguide modes.

A more elegant way to compute the transfer properties makes use of the fact that components in passive isotropic materials are reciprocal, i.e. transmission coefficients for the forward and reverse propagation direction are identical. This implies that the transfer matrix of the receiver region is found directly (by transposition) from that of a transmitter region in which the transmitter waveguides are placed such that both regions are mirrored replicas of each other. The matrix R thus follows from:

$$r_{ij} = a\,U_j(\alpha_i, f), \tag{68}$$

in which $U_j(\alpha_i, f)$ is the diffraction field of the j-th receiver (considered as a virtual transmitter) at the center of the i-th array waveguide, which follows by substitution of

$$U_1^+(y) = U_m(y-y_j) \tag{69}$$

into the integral of equation 6.58a. In the above equation $U_m(y)$ is the modal field of the receiver waveguide and y_j is the position of the *j-th* receiver (= virtual transmitter).

6.6.5 Simulation of dispersion properties

Inspection of equations 6.58a and 6.69 reveals that, apart from a factor $exp(-j\beta_f f)$ which describes the propagation through the free propagation region, the elements of the transmitter and receiver matrices $\boldsymbol{S}$ and $\boldsymbol{R}$ are only weakly dependent on the wavelength (or propagation constant) within the free propagation region. The dispersive properties of the component are determined by the dispersive properties of the array. This greatly facilitates simulation of these properties because the (rather complicated) transmitter and receiver matrices ($\boldsymbol{S}$ and $\boldsymbol{R}$) may be considered constant, the only wavelength dependent matrix being the (very simple) propagation matrix $\boldsymbol{P}$.

6.6.6 Example

Figure 6.28 shows the transmission to the two receiver channels of the polarization splitter discussed in section 6.6.9, as a function of the internal wavelength λ_g within the array waveguides. The wavelengths corresponding to *TE* and *TM* polarization are indicated in the figure. Insertion loss can be interpreted as the level in a channel at the coreesponding wavelength (the wavelengths indicated in the figure as λ_{TE} and λ_{TM}. Cross talk is found as the signal level in a channel at the wavelength of the other. Both insertion loss and cross talk appear to be close to the design values (0.6 and -40 dB respectively).

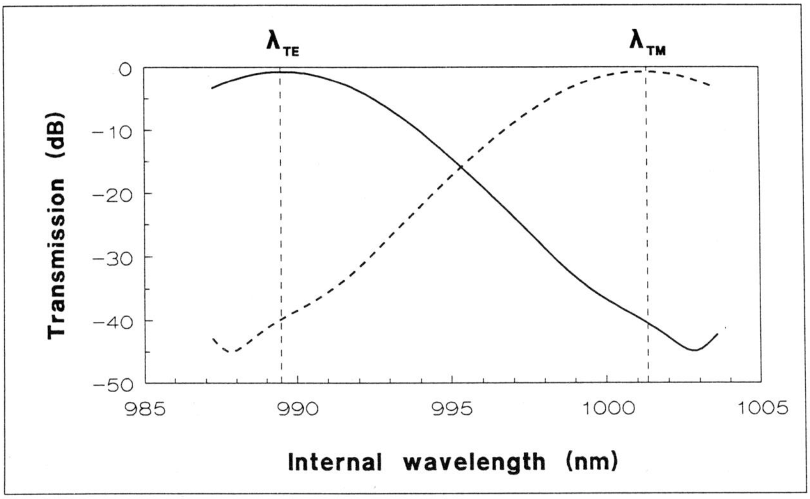

Figure 6.28 *Simulated transmission curves. Wavelengths corresponding to TE- and TM-polarization are indicated.*

6.7 Experiments

The experimental results described in this section are based on components realized according to the initial strategy shown in figure 6.5a. Experiments with the new design approach described in the present thesis, are underway but not yet finished. The most important difference between the two approaches is the absence of chromatic aberration in the new approach. In the initial approach the aberration was corrected for by adapting the receiver positions, as shown in figure 6.30b. Because of the magnitude of this correction, shadowing of one waveguide by the other occurs which causes the cross-talk level to increase. From the viewpoint of performance the most important difference is, therefore, expected to be seen in the cross-talk level. Although the two approaches are essentially similar, the reader should be aware of some difference in the details. Because the new design approach is considered superior, explanation of the initial strategy is restricted to a minimum. Some detail is provided in Appendix 6A.

6.7.1 Polarization splitter

A. Introduction

Several concepts have been reported regarding planar polarization splitters. Bergman et al. [13] proposed a splitter based on intersecting waveguides (device length 25 mm) which showed 1.7 dB insertion loss and 12-15 dB cross talk. Ura et al. [14] demonstrated a focussing grating coupler with 4 dB coupling loss and 14-17 dB cross talk, and a total device area of 1x11 mm^2. Aarnio [15] reported the smallest splitter device so far, which was also based on a grating coupler (device length 5 mm), but with a channel isolation of 9-11 dB. Neyer [16] reported a 3 dB loss figure and 20 dB cross talk for a configuration based on Two Mode Interference (device length 20 mm). Mikami [17] achieved loss and cross-talk values of 1.7 dB and 20 dB respectively using a directional coupler, with 15 mm length. His device requires a tuning circuit for proper operation.

In the following paragraphs it is shown that polarization splitters based on optical phased arrays combine insertion loss and cross-talk figures comparable to the best results reported so far, with considerably smaller device dimensions. Experimental results were reported earlier by Vellekoop and Smit [6,7].

B. Design

The polarization splitter was designed at a wavelength of 633 nm because of the availability of accurate measurement equipment operating at this wavelength. Design was based on 2 μm wide waveguides with a lateral index contrast $\Delta n \simeq 0.02$. These waveguides combine a waveguide attenuation of 2 dB/cm with negligible radiation loss for radii greater than 700 μm.

A waveguide separation of 3 μm (5 μm at centerlines) was chosen. This provides a theoretical channel isolation of 60 dB. The splitter was designed on the basis of an insertion loss L_{max} of 0.5 dB, which occurs, with the chosen waveguide parameters, at a dispersion angle (relative to the optical axis) of 1.5°. At this angle a free propagation length f of 100 μm is required to arrive at a receiver spacing of 5 μm. For the 2 μm waveguides chosen, the $1/e^2$ far-field half angle θ_o amounts to 6°. With $f = 100\ \mu m$ and $W_\alpha = 4\theta_o$ the array counts $n_a = 24$ waveguides. The corresponding cross talk is expected to be lower than - 40 dB.

In the design scheme according to figure 6.5a, the phased array consists of a concentric section, a fan-out and fan-in section for reducing coupling loss, and two adapter sections for connecting the latter sections with the concentric one. The geometry is depicted in more detail in figure 6.29. A straight

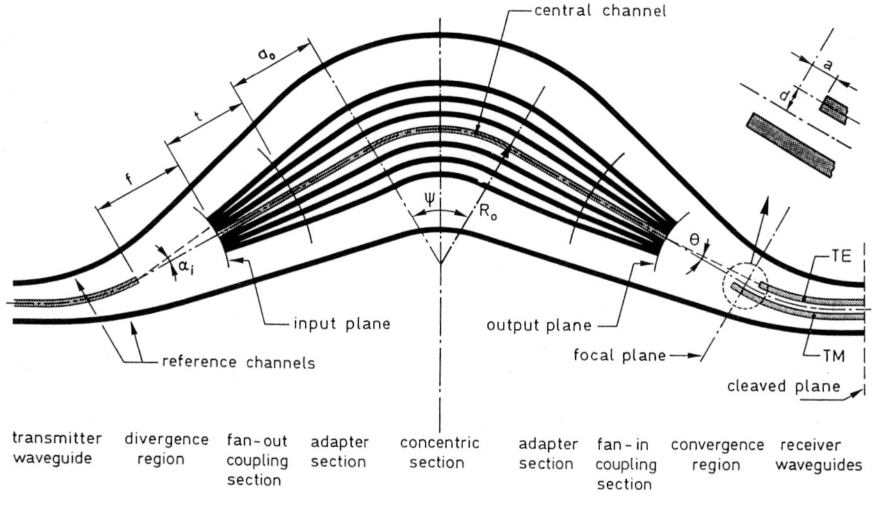

Figure 6.29 Polarization splitter geometry with corresponding design parameters (drawing is not to scale).

section length s (f+t in figure 6.29) of twice the free propagation length was chosen for the array. To avoid radii smaller than 700 µm, which is about the minimum for low-loss bends in this waveguide structure, R_o was chosen 800 µm. The adapter sections contain two circular waveguide sections with slightly different radii, in order to connect the fan-in and fan-out sections smoothly with the concentric section. If too small an adapter central section length a_o is chosen, the radius of one of the circular sections may become smaller than 700 µm. It follows from appendix 6A that this will not happen if $a_o \geq R_o \sin\alpha_{max}$ (in which α_{max} is the value of α_i for the outer array waveguides). This requirement is satisfied by $a_o = 190\ \mu m$.

Once s, f and a_o are determined, the concentric-section sector angle can be determined. For the present waveguide structure the central wavelength $\lambda_c = ½(\lambda_{TE}+\lambda_{TM})$ within the waveguide amounts to 0.406 µm, the relative difference between λ_{TE} and λ_{TM} is approximately 1.3%. With this relative dispersion value the concentric section angle Ψ should equal $\Psi = 1.3$ radians, as follows from Appendix 6A. Corresponding design dimensions, including transmitter and receivers, are $2.5 \times 0.6\ mm^2$.

The dispersive properties of the splitter also apply to wavelength variations, so that a wavelength shift $\Delta\lambda$ will decrease the splitter performance. As long as these variations are within ± 10% of the relative difference $(\lambda_{TM}-\lambda_{TE})/\lambda_c$, the effect will be small. With a relative dispersion of 1.3% the acceptable wavelength fluctuation is in the order of $\pm 10^{-3}$, i.e. approximately ±1 nm.

Table 6.1 summarizes the design parameters and the corresponding predicted properties of the phased array. Figures 6.30a and 6.30b show the experimental device. The two bent waveguides at each side of the phased array and the two straight waveguides are used as reference channels. Loss and cross-talk values were determined by comparison of the receiver channels with the simultaneously excited reference channels.

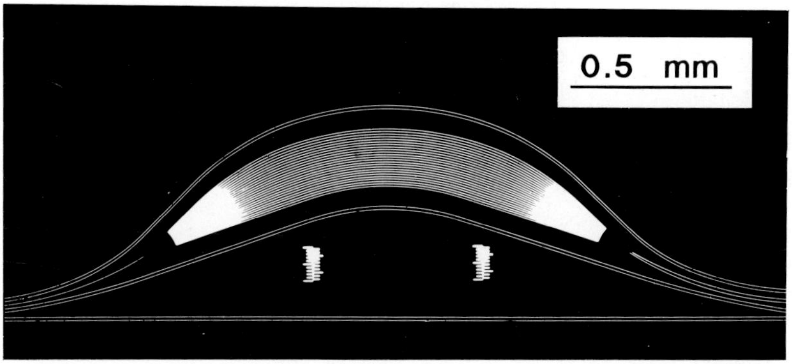

Figure 6.30a *Microscope photograph of the polarization splitter. The two waveguides at each side of the phased array are reference channels.*

Figure 6.30b *Detail of the receiver section showing the aberration correction applied to the two receiver channels.*

vacuum wavelength	λ_o	=	633	nm
TE-wavelength	λ_{TE}	=	0.4032	μm
TM-wavelength	λ_{TM}	=	0.4085	μm
central design wavelength	λ_c	=	0.4058	μm
polarization dispersion	$\Delta\lambda_{pol}/\lambda_c$	=	1.3	%
$1/e^2$ modal half width	w_o	=	1.2	μm
$1/e^2$ far-field half angle	θ_o	=	6.2°	
free propagation length	f	=	100	μm
adapter central length	a_o	=	190	μm
straight section length	s	=	200	μm
central channel radius	R_o	=	800	μm
concentric section angle	Ψ	=	1.3	rad
number of array waveguides	n_a	=	24	
array aperture	$W_{\alpha,max}$	=	26°	
device dimensions			2.5x0.6	mm^2
dispersion angle	$\theta_{TE,TM}$	=	±1.5°	
dispersive displacement	$y_{TE,TM}$	=	±2.5	μm
aberration	$a_{TE,TM}$	=	±3.4	μm
insertion loss	L_{max}	=	0.45	dB
channel cross talk	C	<	- 50	dB

***Table* 6.1** *Summary of design parameters for the polarization splitter.*

C. Simulation

Figure 6.31 shows the power coupled to the TE- and TM-channel as a function of β. The lobes occurring below -40 dB are caused by the finite aperture of the phased array, and agree well with the coarse estimate made based on the spillover. As can be seen from the figure, the predicted insertion loss and channel cross talk at the values of λ corresponding to λ_{TE} and λ_{TM}, are approximately -0.5 dB and - 50 dB.

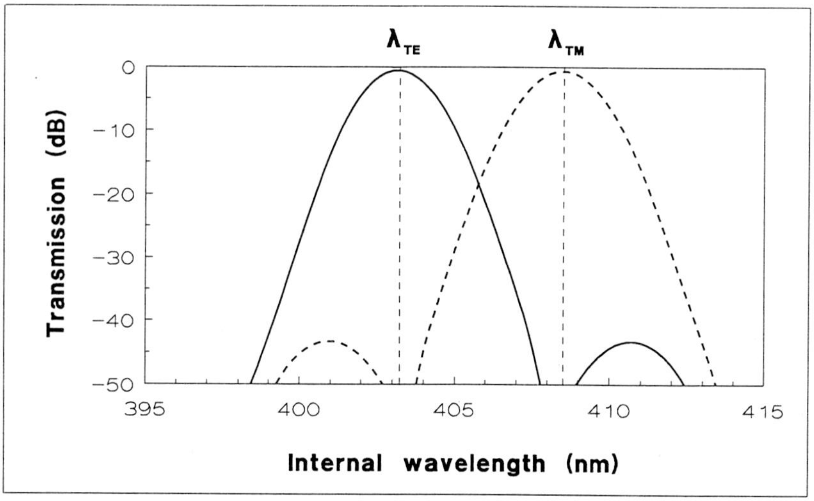

__Figure 6.31__ The numerically simulated power transfer for the TE and the TM channel, as a function of the internal wavelength λ_g. The design values for λ_{TE} and λ_{TM} are indicated.

D. Results

The experimental device was investigated by selectively coupling *TE*- or *TM*-polarized light from an He-Ne gas laser into the planar waveguides with a prism coupler, as described in chapter 8. The device was cleaved at the position indicated in figure 6.29. The output end face of the device is projected onto a CCD video camera with a microscope objective. The signal from the camera is digitized and processed by a computer. Figure 6.32 shows photographs of the camera observations for the two polarizations. An intensity scan over the different channels is shown in figure 6.33.

Insertion loss can be determined by comparing the intensity of the receiver channel with the intensities of the reference channels as shown in figure 6.32. Table 6.2 lists the values measured for four different devices. The values measured for devices 3 and 4 are less representative than the others due to a

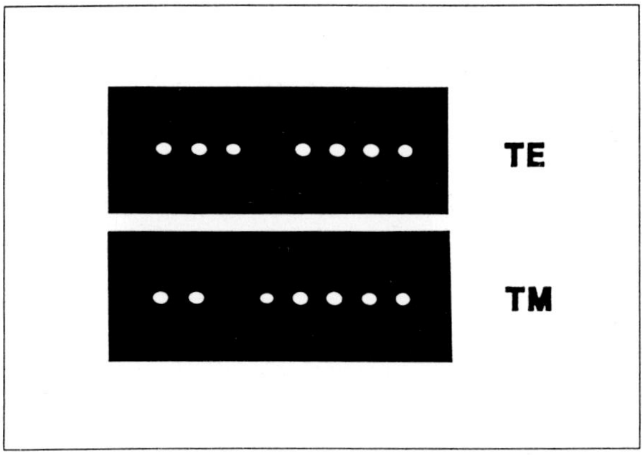

Figure 6.32 *Camera observations of the channels at the cleaved end face of the device (as indicated in figure 6.29) for TE- and TM-polarization. The two channels at the left and the four channels at the right are reference channels.*

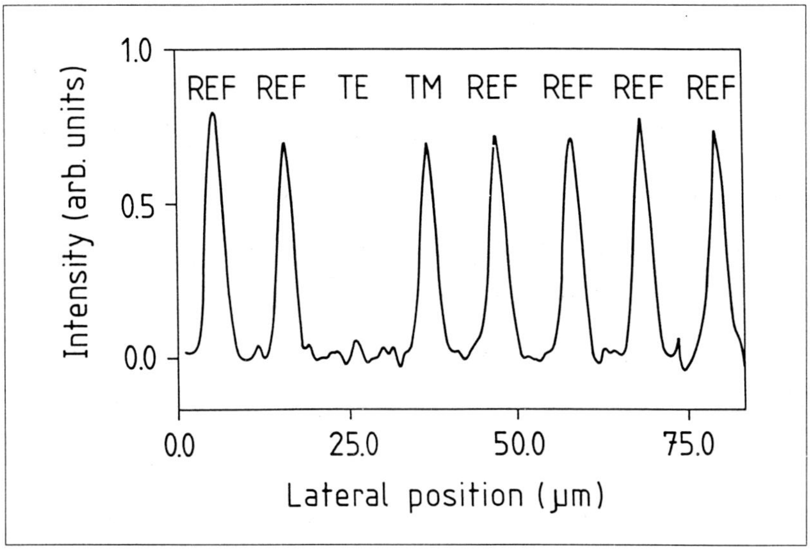

Figure 6.33 Intensity scan of the TM-polarized response.

Nr.	mismatch $\Delta\lambda/\lambda$ (%)	TE Loss (dB)	TE Cross talk (dB)	TM Loss (dB)	TM Cross talk (dB)
1	0.03	2.0	−23.1	2.0	−21.3
2	0.05	1.7	−18.7	0.3	−20.5
3	0.12	1.5	−19.1	0.1	−13.4
4	0.12	2.1	−21.1	0.3	−17.3

Table 6.2 Measurement results of four different devices. The measured wavelength mismatch of the experimental device relative to the design value, is indicated in the second column.

poor match of the measured values of λ_{TE} and λ_{TM} to the design values. It can be seen that an insertion loss close to the theoretically predicted value of 0.5 dB is obtained for the TM channel (device 2). The lowest values listed in the table are not considered significant: measurement errors are estimated to be within ±0.5 dB. The *TE*-polarized channel exhibits a greater insertion loss than the *TM*-polarized one. This may have been caused by shadowing effects from the *TM*-polarized receiver which extends to within the field-of-view of the *TE*-receiver due to the aberration corrections, as can be seen from figure 6.30b.

Channel isolation is determined by comparing the transmitted signal intensity for a single polarization with the signal level in the other channel after removing a 20 dB optical attenuator. The cross-talk values of 19-20 dB (devices 1 and 2) are worse than the theoretical predictions, but comparable to best results published so far.

E. Conclusions

A polarization splitter based on an optical phased array was realized with insertion loss and cross-talk figures of 2 and -19 dB for *TE*-polarization, and 0.5 and -20 dB for *TM*-polarization. Device dimensions are $2.5 \times 0.6\ mm^2$. This performance is comparable with the best results reported so far, but realized with a considerably smaller device size.

6.7.2 Polarization-independent wavelength demultiplexer

A. Introduction

Various configurations of planar (de)multiplexers have been reported based on optical interference filters [18-20], wavelength selective coupling [21-23] and optical diffraction gratings [24-26]. Verbeek et al. [18] reported a weakly polarization-dependent four-channel Mach-Zehnder demultiplexer (device length 33 mm) with a channel spacing of 7.7 nm, 2.6 dB loss and 16 dB channel cross talk. Rottman et al. [21] realized a polarization-independent dual-channel (de)multiplexer with 30 nm channel spacing, 3 dB loss and 26-40 dB cross talk, based on two-mode interference (TMI) and with a device length of 12 mm. Imoto et al. [22] proposed and demonstrated a dual-channel directional coupler (de)multiplexer with 100 nm channel spacing, 5 dB loss and 20 dB cross talk. The configuration was designed for a specific polarization and had a length of 10 mm. Suhara et al. [24] demonstrated a five-channel demultiplexer for multimode systems with 30 nm channel spacing using a chirped grating. The hybrid component, with dimensions 20x13 mm^2, showed 9.5 dB loss and 21.5 dB cross talk. Non-planar grating components using GRIN lenses [27,28] achieve better performance. Such devices must be assembled, however, with great precision and are therefore less attractive.

Polarization-independent wavelength demultiplexers based on optical phased arrays combine insertion loss and cross-talk figures comparable to the best results reported so far, but with considerably smaller device dimensions.

Experimental results were reported earlier by Vellekoop and Smit [8,9]. A short description will be given below.

B. Design

A single mode AlGaAs laser, which is thermally tunable between 775-785 nm, was employed to measure the device, thus fixing the design central wavelength at λ_o = 780 nm. The design aim was to achieve a polarization-independent 4-channel wavelength demultiplexer with a wavelength spacing of 1.7 nm. The same (bimodal) waveguide structure (lateral waveguide width $w=2\ \mu m$, optical contrast $\Delta n \simeq 0.02$), which was succesfully employed in the polarization splitter was chosen for the demultiplexer. Due to the slightly longer wavelength (780 nm instead of 633 nm), the minimal bending radius is increased to 1 mm. The central design wavelength λ_c (0.5043 μm), the $1/e^2$ modal half width w_o (1.34 μm) and the relative polarization dispersion $\Delta\lambda_{pol}/\lambda_c$ (1.37%) are fixed by the choice of the waveguide geometry.

Receiver waveguide separation was chosen to be similar to the polarization splitter: 5 μm at centerlines. Due to the higher operation wavelength the predicted channel cross-talk level was higher: -45 dB. This is, however, considered to be sufficiently low. The demultiplexer was made polarization independent by choosing the free propagation length according to equation 6.51, which leads (with $d=w=2\ \mu m$ and $i=2$) to a value $f=123\ \mu m$. With a dispersive displacement of $1\frac{1}{2} x 5 = 7.5\ \mu m$ for the outer channels the dispersion angle θ amounts to approximately 3.5°. This will produce an insertion loss of 2 dB for the outer channels. Channel cross-talk considerations are analogous to the polarization splitter design. The receiver-waveguide spacing of 5 µm and a number $N_a = 31$ array waveguides is calculated to be sufficient to achieve a channel isolation better than -40 dB.

vacuum wavelength	λ_o	=	780	nm
central design wavelength *(TE)*	λ_c	=	0.5043	µm
relative channel spacing	$\Delta\lambda/\lambda$	=	0.22	%
polarization dispersion	$\Delta\lambda_{pol}/\lambda_c$	=	1.37	%
$1/e^2$ modal half width	w_o	=	1.34	µm
$1/e^2$ far-field half angle	θ_o	=	6.9°	
free propagation length	f	=	123	µm
adapter central length	a_o	=	300	µm
straight section length	s	=	583	µm
central channel radius	R_o	=	1200µm	
concentric section angle	Ψ	=	180°	
number of array waveguides	n_a	=	31	
array aperture	$W_{\alpha,max}$	=	28°	
device dimensions			4.5x3.2	mm^2
dispersion displacement	$\theta_{1,4}$	=	±7.5	µm
	$\theta_{2,3}$	=	±2.5	µm
aberration	$a_{1,4}$	=	±4.5	µm
	$a_{2,3}$	=	±1.5	µm
insertion loss	$L_{1,4}$	=	2.3	dB
	$L_{2,3}$	=	0.3	dB
channel cross talk	C	<	-40	dB

***Table 6.3** Summary of design parameters for the wavelength demultiplexer.*

If the concentric section angle Ψ is restricted to 180° and the adapter section reference length a_o is chosen to be 300 μm, the straight section length for obtaining the required dispersion is found to be 583 μm. Corresponding device dimensions are $4.5x3.2\ mm^2$.

If the demultiplexer is designed for *TE*-polarization, the aberration-correction of the receiver channels, as described in the appendix at the end of this chapter, only applies to the *TE*-polarization. The aberration for the *TM*-field is considerable. Substituting $\Delta\lambda_{pol}/\lambda_c = 1.37\%$ in equation 84 of the appendix an aberration a_{TM} of 19 μm is found. Compared to the focal depth $d_f = \pi w_o^2/\lambda_c \simeq 11$ *μm* (this being the distance for which the increase in beam width is within a factor √2), it is to be expected that the demultiplexer properties for *TM*-polarization will be inferior to those for *TE*-polarization.

Polarization performance can be improved by placing the receiver waveguides so that the focal mismatch is equal for both polarizations. This is achieved by correcting the receiver position by the average of the aberration for the two polarizations. We decided to optimize the aberration correction for a single polarization *(TE)* because the demonstration of demultiplexer operation was considered to be a first priority. The aberration-free geometry described in the previous sections was developed based on the results of this experiment.

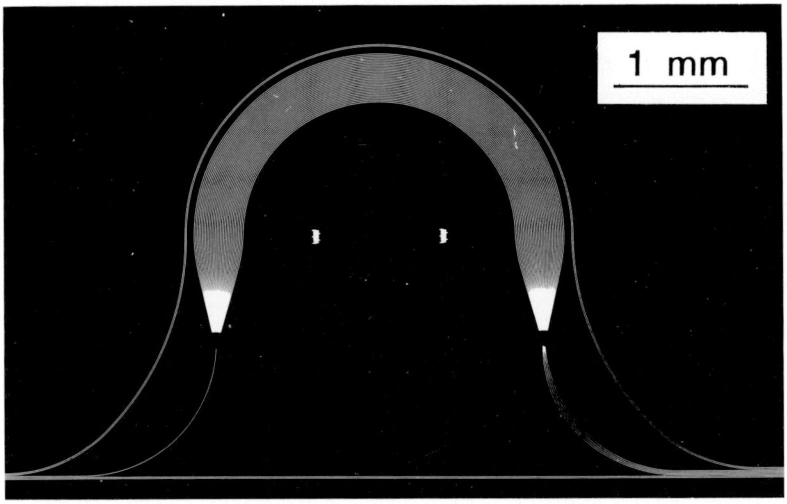

Figure 6.34a *Mask pattern of the demultiplexer. The upper two bent waveguides and the lower two straight waveguides are reference channels.*

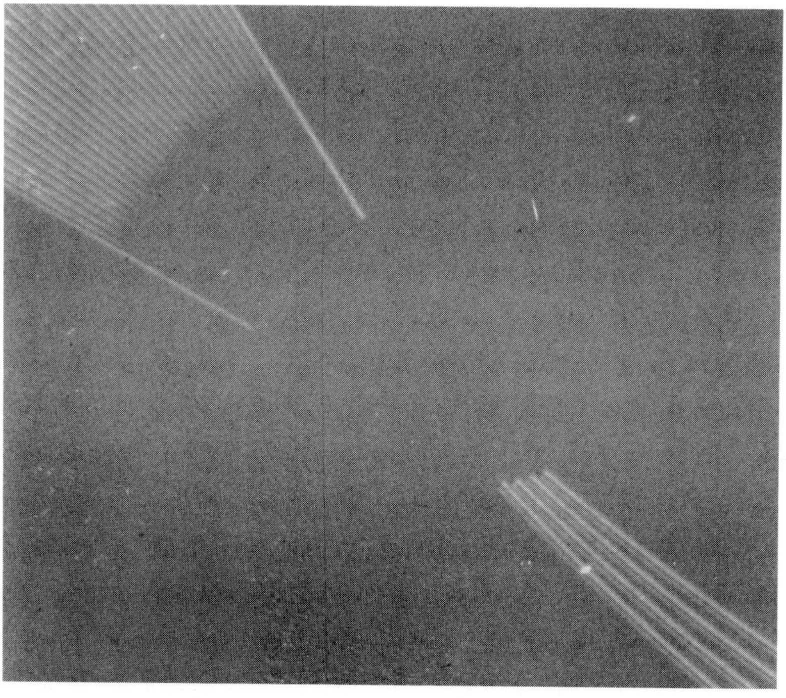

Figure 6.34b *Microscope photograph of the receiver section.*

Design parameters and corresponding theoretical properties are summarized in table 6.3. Figure 6.34 shows a photograph of the chromium mask for the demultiplexer. The two bent waveguides at the upper side and the two straight waveguides at the lower side of the phased array are used as reference channels. Figure 6.34b shows a photograph of the receiver section on the chip.

C. Simulation

Simulation results are shown in figure 6.35. The simulation predicts an insertion loss below 2.5 dB and a cross talk of -40 dB, which is 5 dB higher than the design value due to inclusion of aberration and truncation effects (as a result of the finite aperture width).

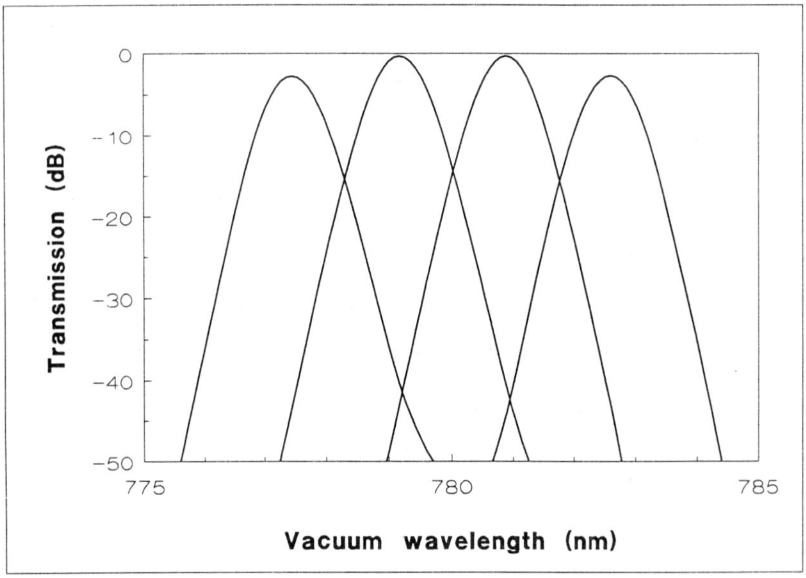

Figure 6.35 The numerically simulated power transfer as a function of the vacuum wavelength λ.

As the demultiplexer is designed for a specific wavelength, production tolerances or thermal effects may shift the wavelength filter characteristics of the demultiplexer, thus reducing demultiplexer performance. Variations within ±10 percent of the relative channel spacing $\Delta\lambda/\lambda_c$ have negligible effect. With the above values, the acceptable relative propagation constant fluctuation lies within ±10% of 1.7nm at a wavelength of 780 nm, i.e. within $\pm 2\ 10^{-4}$, corresponding to a maximum wavelength fluctuation $\Delta\lambda \leq \pm 0.17\ nm$.

D. Results

The experimental device was investigated by coupling light into the device from an AlGaAs semiconductor laser, operating around 780 nm, with a prism coupler. Input and reference channels were excited simultaneously with a broad beam. Transmission loss was determined by comparing the output intensity to that of the reference channels. Maximum measurement errors, as determined from comparison of identical (straight) waveguides, were within ±0.5 dB. A thermoelectric Peltier module was used to tune the semiconductor laser wavelength between 775 nm and 785 nm with a temperature coefficient $d\lambda/dT \simeq 0.25\ nm/K$. The cleaved end face of the device, as indicated in figure 6.28 (for the polarization splitter), was projected onto a CCD video camera with a microscope objective. The camera signal is digitized and processed by a computer.

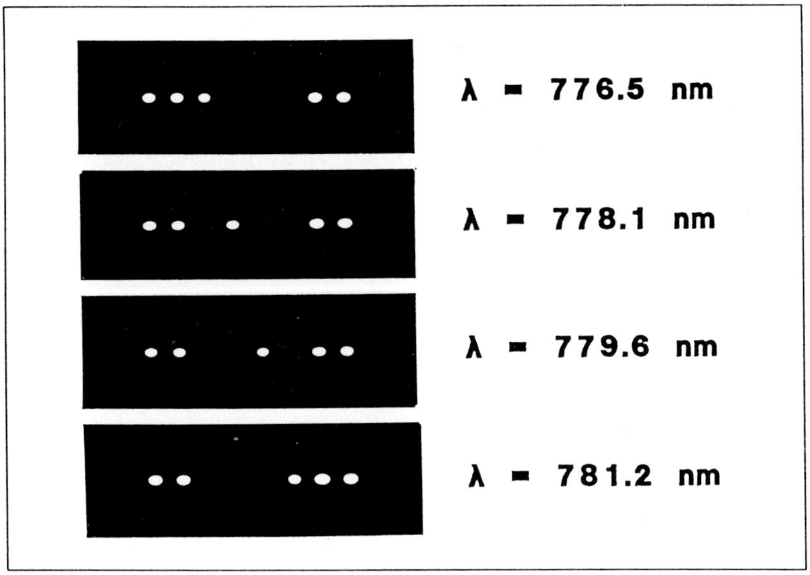

Figure 6.36 *Camera observations of the channels at the cleaved end face of the device for the four consecutive channel wavelengths. The two channels at the left and the right are straight and curved reference channels, respectively.*

Figure 6.36 shows the intensity pattern observed at the cleaved end face at different wavelengths. Insertion loss is determined by comparing the intensity from the receiver channel with that from the reference channels. For the *TE*-polarization an insertion loss of 0.6 dB was found for the central channels and 1.2 dB for the outer channels; for the *TM*-polarization the insertion loss measured 2.1 dB for the central and 3.2 dB for the outer channels.

TE insertion loss of 1.2 dB for the outer channels is 1 dB lower than the theoretically expected value. This may be explained as follows. The limited resolution of the optical lithography results in a partial filling of the inter-waveguide gap close to both ends of the array, as can be seen from the photographs in figures 6.34a and 6.34b. Due to this filling, the actual focal length will become longer than the design value, and the actual length of the fan-in-section will be reduced by the same amount. The corresponding decrease of the actual dispersion angles will lead to a lower insertion loss.

The length over which the gaps are filled has been experimentally determined to be approximately 70 μm. With a corrected free propagation length $f=193$ μm the actual dispersion angle for the outer channels becomes 2.2°. For this geometry we find an angular insertion loss $L_\theta \simeq 0.9\ dB$ for the outer channels (equation 6.19) and a central insertion loss $L_o \simeq 0.4\ dB$ (equation 6.22), so that a total loss of 1.3 dB is expected. The measured insertion loss of 1.2 dB for the outer channels appears to be in good agreement with this corrected value. Waveguide propagation loss are not included in these results which were obtained by comparing the intensities in the output channels to those in the reference channels. This loss is estimated from experimental results to amount to 1.3 dB.

Channel isolation is determined by comparing the intensities in the receiver channels, after removing a 20 dB optical attenuator, to that of the reference channels, and measured 15.4-29.7 dB for the *TE*-polarization and 13.4-22.2 dB for the *TM*-polarization. Minimum cross talk appears to occur at wavelength spacings (1.55 nm) slightly less than the design value (1.7 nm).

Table 6.4 lists the results measured for the demultiplexer. The laser wavelength was measured by coupling part of the input beam to a monochromator with ±0.1 nm resolution. TM polarized results appear to be approximately 2 dB poorer than the TE-results, both for insertion loss and cross talk, as might be expected from the focal mismatch due to aberration.

Channel	Wave-length (nm)	TE transmitted power (dB)			
		1	2	3	4
1	776.5	-1.2	-19.0	-21.3	-25.7
2	778.1	-19.4	-0.6	-20.0	-20.6
3	779.6	-17.5	-16.2	-0.8	-15.4
4	781.2	-29.7	-21.7	-16.0	-1.2

Channel	Wave-length (nm)	TM transmitted power (dB)			
		1	2	3	4
1	776.5	-2.8	-17.0	-21.5	-22.2
2	778.1	-17.6	-2.1	-16.5	-20.0
3	779.6	-20.4	-17.4	-3.2	-15.9
4	781.2	-18.8	-19.1	-13.4	-3.0

Table 6.4 Measurement results of the demultiplexer for TE-polarization (upper table) and TM-polarization (lower table).

E. Conclusions

The experimental demultiplexer measured an insertion loss (excluding 1.3 dB waveguide propagation loss) of 0.6 dB for the central channels and 1.2 dB for the outer channels for *TE*-polarization, and 2.1 dB and 3.2 dB respectively for *TM*-polarization. Channel isolation was 15.4-29.7 dB for *TE*- and 13.4-22.2 dB for *TM*-polarization relative to the reference channels. Channel wavelength spacing is 1.55 nm. These properties are comparable to the best results reported so far, but realized with a considerably smaller device size.

6.8 Discussion and conclusions

Optical phased arrays combine focussing and dispersive properties on a small device area. They provide an alternative for curved planar gratings and can be realized simultaneously with the waveguide pattern thus avoiding the need for holographic or direct-write electron-beam lithography. Application as a polarization splitter and as a wavelength demultiplexer has been demonstrated, application as a wavelength filter is straightforward.

The experimental results reported were based on a design strategy which introduces chromatic aberration at the receiver side. A better approach has been described which avoids chromatic aberration and simplifies component design. A design strategy has been developed with which optimal design can be performed using a set of normalized graphs.

APPENDIX Design of a phased array based on a concentric section

A. Phase transfer

The main difference between a phased array based on a concentric section, as depicted in figures 6.5a and 6.29, and the newer scheme of figure 6.5b, as described in the present thesis, is that the straight section length is constant in the older scheme, while the radius of the concentric sections is used to achieve the required phase transfer. An adapter section is required for a smooth connection between the straight and the concentric waveguide sections. This section contributes to the phase transfer and is responsible for the aberration which was seen to degrade the TM-performance of the demultiplexer discussed in section 6.7.2.

Employing the array geometry as depicted in figure 1, the phase transfer of the i-th channel relative to the reference channel follows as

$$\Phi_i = -\beta\,(l_i - l_o) = -\beta\,\{2(a_i - a_o) + \Psi(R_i - R_o)\}, \tag{70}$$

in which $\Psi\,(R_i - R_o)$ is the difference in length between the i-th and the central channel of the concentric section and $2(a_i - a_o)$ is the same difference for the two adapter sections. To obtain focussing action the radii R_i are chosen to be equidistant, and are then rounded off to the value for which the relative phase transfer Φ_i equals an integer multiple of 2π.

The individual waveguides of the fan-in and fan-out coupling section are described by their starting angle α_i relative to the reference angle α_o of the array for which the curvature of the adapter section equals zero (see figure 6.29). The i-th waveguide of the fan-in section is smoothly connected to the concentric section by an arc over an angle α_i with a radius

$$\rho_i = \frac{\alpha_o + s(1 - \cos\alpha_i)}{\sin\alpha_i}, \tag{71}$$

in which a_o is the reference length of the adapter section, s is the straight section length, and f is the free propagation length. The radius of the corresponding channel in the concentric section is

$$R_i = R_o + 2(s + ½a_o)\tan ½\alpha_i, \tag{72}$$

R_o being the radius of the reference channel. The total relative length $l_i - l_o$ of the i-th channel thus becomes :

$$l_i - l_o = \Psi(R_i - R_o) + 2(\alpha_i\rho_i - a_o). \tag{73}$$

To meet the phase transfer requirement for the i-th channel, R_i must be altered by a small amount dR_i (to be calculated later). This change dR_i introduces a connection error between the adapter section and the concentric section. The error can be prevented by dividing the adapter section into two circular waveguide sections, each over an angle $½\alpha_i$, with radii:

$$\rho_{i1} = \rho_i - \Delta\rho_i, \tag{74a}$$

$$\rho_{i2} = \rho_i + \Delta\rho_i - dR_i, \tag{74b}$$

in which

$$\Delta\rho_i = \frac{dr_i}{2 - \sqrt{(2 + 2\cos\alpha_i)}}, \qquad (\alpha_i \neq 0). \tag{75}$$

If $\alpha_i = 0$, i.e. for the reference element, $dR_i = 0$ and the adapter waveguide becomes straight. The length a_i of the i-th channel in the adapter follows as

$$a_i = \alpha_i(\rho_i - \tfrac{1}{2}\, dR_i) \tag{76}$$

so that the corresponding total relative length becomes :

$$l_i - l_o = \Psi(R_i - R_o) + 2(\alpha_i\rho_i - a_o) - dR_i(\Psi + \alpha_i). \tag{77}$$

To obtain focussing action the dR_i have to be chosen such that $(l_i - l_o)$ equals an integer number of wavelengths. If the solution is chosen closest to zero for the dR_i , R_i will become nonlinear in α_i , as can be seen from equation 6.72. The tangent of α_i being described by $\alpha_i + \frac{1}{3}\alpha_i^3 + O(\alpha^5)$, this nonlinearity can be avoided by choosing the value closest to $\frac{1}{12}(s+\tfrac{1}{2}a_o)\,\alpha_i^3$ for dR_i[1].

B. Dispersion and aberration

If we substitute the expressions for $R_i - R_o$ (equation 6.72), ρ_i (equation 6.72), and dR_i (the series expansion $\frac{1}{12}(s+\tfrac{1}{2}a_o)\,\alpha_i^3$ into the expression for the relative channel length $l_i - l_o$ (equation 6.72) we can develop the resulting expression into into the following Taylor series in α_i

$$(l_i - l_o) = \psi(s+\tfrac{1}{2}a_o)\,\alpha_i + (s+a_o/3)\,\alpha_i^2 + \frac{1}{6}(s+\tfrac{1}{2}a_o)\,\alpha_i^3 + O(\alpha_i^4)\,. \tag{78}$$

Through a proper choice of the R_i's the array is designed such that $\Phi_i(\beta_c)=0$ for all i. Consequently

$$\Phi_i(\beta) = -(\beta - \beta_o)\,(l_i - l_o). \tag{79}$$

---

1) This choice was considered later as a less fortunate one. A better choice would have been to choose the values of dR_i such that the element lengths l_i are linear in α_i. This would have eliminated the aberration.

The phase transfer function $\Phi_i(\beta)$ thus contains linear, quadratic and higher order terms in α_i. The linear term describes the dispersion of the array. The tilting angle $d\theta$ of the outgoing wave front is determined by the tangent of the phase difference between two adjacent channels, expressed as a fraction of the wavelength $(\Delta\Phi/\beta)$, divided by the distance $f\,\Delta\alpha$ between the channels according to :

$$\frac{d\theta}{\delta\beta} \simeq \frac{\frac{d}{d\beta}(\Phi_{i+1} - \Phi_i)}{\beta f(\alpha_{i+1} - \alpha_i)} = \frac{-1}{\beta^2 f}\frac{d\Phi}{d\alpha} . \tag{80}$$

At the receiver end this angular tilt will cause a lateral displacement of the focal point :

$$d = f\frac{d\theta}{d\beta}\Delta\beta = \frac{\Delta\beta}{\beta_c}\Psi'(s+\tfrac{1}{2}\alpha_o) = -\frac{\Delta\lambda}{\lambda_c}\Psi'(s+\tfrac{1}{2}\alpha_o) . \tag{81}$$

At the central propagation constant β_c, the wave-front curvature in the array aperture is *1/f*. the second order term of equation 6.78 introduces an additional phase difference between adjacent channels proportional to the square of the angular coordinate of the i-th channel (α_i) for a variation of the propagation constant. This phase difference leads to an additional wave-front curvature Δc, described by the second derivative of Φ. It can be inferred that this extra curvature is described by :

$$\frac{dc}{d\beta} = \frac{1}{f^2}\frac{d^2\Phi/d\alpha^2}{\beta^2} . \tag{82}$$

The corresponding change in the focal length is :

$$\frac{df}{d\beta} = \frac{d(\frac{1}{c})}{d\beta} = -f^2\frac{dc}{d\beta} = \frac{1}{\beta^2}\frac{d^2\Phi}{d\alpha^2} . \tag{83}$$

The aberration *a* is thus found as :

$$a = \frac{df}{d\beta}\,\Delta\beta = 2\,\frac{\Delta\beta}{\beta_c}(s+a_o/3) = -2\,\frac{\Delta\lambda}{\lambda_c}\,(s+a_o/3). \tag{84}$$

For small values of a_o the ratio d/a approaches ½ Ψ. For small values of Ψ the aberration may therefore become considerably greater than the dispersion.

References.

6.1 D.B. Anderson, R.L. Davis, J.T. Boyd, and R.R. August, "Comparison of optical-waveguide lens technologies", IEEE J. Quantum Electron., Vol. 13, pp. 275-282, 1977.

6.2 G. Hakatoshi, H. Inouie, K. Naito, S. Umegaki, and S. Tanaka, "Optical waveguide lenses", Opt. Acta, Vol. 26, pp. 961-968, 1979.

6.3 T. Suhara and H. Nishihara, "Integrated optics components and devices using periodic structures", IEEE J. Quantum Electron., Vol. 22, pp. 845-867, 1986.

6.4 P.C.A. Hammes, G.J.M. Krijnen, A. Driesen, and M.K. Smit, "High efficiency small-size planar hyperbolical lenses", Electron. Lett., Vol. 26, pp. 1064-1065, 1990.

6.5 M.K. Smit, "A new focussing and dispersive planar component based on an optical phased array", Electron. Lett., Vol. 24, pp. 385-386, 1988.

6.6 A.R. Vellekoop and M.K. Smit, "Low-loss planar optical polarisation splitter with small dimensions", Electron. Lett., Vol. 25, pp.946-947, 1989.

6.7 A.R. Vellekoop and M.K. Smit, "A polarization independent planar wavelength demultiplexer with small dimensions", Procs. Eur. Conf. on Opt. Integr. Systems at Amsterdam, The Netherlands, September 25-28, 1989, paper D3.

6.8 A.R. Vellekoop and M.K. Smit, "A small size polarization splitter based on a planar optical phased array", J. Lightwave Technol., Vol. 8, pp. 118-124, 1990.

6.9 A.R. Vellekoop and M.K. Smit, "A polarisation-independent four-channel wavelength demultiplexer based on a planar optical phased array", J. Lightw. Technol., Vol.9, pp. 310-314, 1991.

6.10 H. Takahashi, S. Suzuki, K. Kato, and I. Nishi, "Arrayed-waveguide grating for wavelength division multi/demultiplexer with nanometre resolution", Electron. Lett., Vol. 26, pp.87-88, 1990.

6.11 H. Takahashi, S. Suzuki, Isao Nishi, "Multi/demultiplexer for nanometer-spacing WDM using arrayed-waveguide grating", Integrated Photonics Research Topical Meeting, April 9-11, 1991, Monterey, California, USA, Technical Digest Series, Vol. 8, Postdeadline papers, paper PD1, 1991.

6.12 E.E. Bergmann, L. McCaughan, and J.E. Watson, "Coupling of intersecting Ti:LiNbO3 diffused waveguides", Appl. Opt., Vol. 23, pp. 3000-3003, 1984.

6.13 S. Ura, H. Sunagawa, T. Suhara and H. Nishihara, "Focussing grating couplers for polarization detection", J. Lightwave Technol., Vol. 6, pp. 1028-33, 1988.

6.14 J. Aarnio, "An integrated-optic polarization splitter on silicon substrate", Eur. Conf. on Opt. Commun. ECOC '88, Brighton UK, 11-15 Sept. 1988, pp. 37-40, 1988.

6.15 A. Neyer and D. Dang, "Guided-wave polarization splitter based on two-mode-interference", Conf. on Integrated and Guide-Wave Optics, March 28-30, 1988, Santa Fe, New Mexico, Tech. Digest series, Vol. 5, pp.107-110, 1988.

6.16 O. Mikami, "LiNbO3 coupled-waveguide TE/TM mode splitter", Appl. Phys. Lett., Vol. 36, pp. 491-493, 1980.

6.17 H.G. Unger, "Planar optical waveguides and fibres", Oxford : Clarendon press, 1977.

6.18 B.H. Verbeek, C.H. Henry, N.A. Olsson, K.J. Orlowsky, R.F. Kazarinov and B.H. Johnson, "Integrated four-channel Mach-Zehnder multi/demultiplexer fabricated with phosphorous doped SiO_2 waveguides on Si", J. Lightwave Technol., Vol. 6, pp. 1011-1015, 1988.

6.19 K. Oda, N. Takato, H. Toba and K. Nosu, "A wide-band guided-wave periodic multi/demultiplexer with a ring resonator for optical FDM transmission systems", J. Lightwave Technol., Vol. 6, pp. 1016-1023, 1988.

6.20 H. Toba, K. Oda, N. Takato, and K. Nosu, "5 GHz-spaced, eight-channel, guided-wave tunable multi/demultiplexer for optical FDM transmission systems", Electron. Lett., Vol. 23, pp. 788-789, 1988.

6.21 F. Rottmann, A. Neyer, W. Mevenkamp and E. Voges, "Integrated-optic wavelength multiplexers on lithium niobate based on two-mode interference", J. Lightwave Technol., Vol. 6, pp. 946-952, 1988.

6.22 K. Imoto, H. Sano and M. Miyazaki, "Guided-wave multi/demultiplexers with high stopband rejection", Appl. Opt., Vol. 26, pp. 4214-4219, 1987.

6.23 R.C. Alferness and R.V. Schmidt, "Tunable optical waveguide directional filter", Appl. Phys. Lett., Vol. 33, pp. 161-163, 1978.

6.24 T. Suhara, H. Nishihara, J. Koyama, J. Viljanen and M. Leppihalme, "Integrated-optic wavelength multi- and demultiplexers using a chirped grating and an ion-exchanged waveguide", Proc. First European Conference on Integrated Optics ECIO, Sept. 1981, London, UK, post-deadline paper, pp. 12-14, 1981

6.25 R.C. Alferness, L.L. Buhl, M.J.R. Martyak, M.D. Divino, C.H. Joyner and A.G. Dentai, "Narrowband GaInAsP/InP waveguide grating-folded directional coupler multiplexer/demultiplexer", Electron. Lett., Vol. 24, pp. 150-151, 1988.

6.26 Y. Kokubun, T. Baba, H. Watanabe and K. Iga, "Monolithic integration of ARROW-type demultiplexer and photodetector", Proc. European Conference on Optical Communication ECOC'88, Brighton UK, pp 231-234, 1988.

6.27 J. Lipson, W.J. Minford, E.J. Murphy, T.C. Rice, R.A. Linke and G.T. Harvey, "A six-channel wavelength multiplexer and demultiplexer for single mode systems", J. Lightwave Technol., Vol. 3, pp. 1159-1162, 1985.

6.28 M. Seki, K. Kobayashi, Y. Odagiri, M. Shikada, T. Tanigawa and R. Ishikawa, "20-channel micro-optic grating demultiplexer for 1.1-1.6 μm band using a small focussing parameter graded-index rod lens", Electron. Lett., Vol. 18, pp. 257-258, 1982.

Chapter 7

Measurement technique Ellipsometry

Ellipsometry is a widespread technique for measuring the thickness and the refractive index of transparent films on semiconductor substrates. The accuracy of the method is strongly dependent on the optical thickness of the film and the condition of both film and substrate, which complicates the interpretation of measurement results. In this chapter, the accuracy of the method is analyzed using a bounded-error estimation method. Further, a method is developed and tested for estimating both random and systematic errors from a series of measurement data.

7.1 Introduction

Ellipsometry is a quick and nondestructive method for measuring optical film or substrate properties [1,2]. The method is based on measuring the complex ratio of the reflection coefficients of a (film-covered) substrate for light polarized parallel to and normal to the incidence plane, respectively. The complex refractive index of the substrate can be determined from this ratio or, for a film-covered substrate, the film thickness and the refractive index, assuming the substrate index is known. If the latter is not known, or the refractive index has a non-negligible imaginary element, additional measurements are necessary to determine the unknown parameters.

Additional measurements can also be used to increase the measurement accuracy. They can be obtained by measuring at multiple angles of incidence (MAI). MAI measurements find wide application. Accuracy analysis of MAI-measurements has been reported employing Least-Squares Estimation (LSE) methods [3,4] and parameter correlation analysis [5]. Another way to increase the number of independent measurements is by measuring at different wavelengths (MW). MW-measurements have the difficulty that most materials are dispersive; each additional wavelength introduces additional unknown refractive indices. This increase in the total number of unknowns can be reduced if a dispersion equation employing a small number of unknown constants is available [6,7]. Bu-Abbud et al. [8] and Snyder et al. [9] showed that a combination of MAI and MW ellipsometry (MAW) can yield considerably more information than either of the methods alone.

To take full advantage of the methods discussed above, measurement equipment operating along a continuously variable incidence angle and wavelength is required. Such equipment is available only in a limited number of places. Humlícek [4] showed that MAI ellipsometry involving a limited number of incidence angles may yield quite satisfactory results. In this chapter the accuracy which can be obtained using commercially available, inexpensive ellipsometry equipment will be investigated. The analysis is tailored to the wavelength (633 nm) and the three incidence angles (30°, 50°, and 70°) available on a Gaertner 645-AK ellipsometer. Conclusions, however, are not restricted to this specific instrument. Finally, a method for estimating the magnitude of random and systematic errors will be discussed. The results were recently published [10-12].

7.2 Measurement method

Figure 7.1 illustrates the measurement setup. A linearly polarized laser beam, which is fed through a quarter-wave plate transforming it into a circularly polarized beam, is incident to the film-covered substrate with an adjustable incidence angle θ. The polarization state of the beam is altered on reflection. The polarization state after reflection contains information about the film; it is measured with a rotating analyzer.

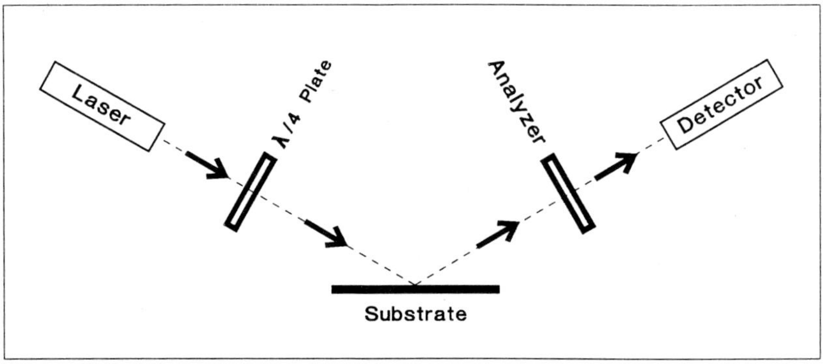

Figure 7.1 Ellipsometer configuration.

Optical film and substrate properties are calculated from the measured complex ratio

$$\rho = \rho^{p}/\rho^{s}, \tag{7.1}$$

of the reflection coefficients ρ^{p} and ρ^{s} for light polarized parallel and normal to the plane of incidence, respectively. ρ is usually represented by the parameters Δ and Ψ, which are defined as:

$$\Delta = arg(\rho), \tag{7.2a}$$

$$\Psi = arctan(|\rho|). \tag{7.2b}$$

For a substrate covered with a single film, ρ^{p} and ρ^{s} follow from:

$$\rho^{p,s} = \left[\frac{r_u^{p,s} + r_l^{p,s} e^{-2j\delta}}{1 + r_u^{p,s} r_l^{p,s} e^{-2j\delta}} \right], \tag{7.3}$$

in which r_u^{p}, r_u^{s}, r_l^{p}, and r_l^{s} are the reflections at the upper (film-air) and the lower (film-substrate) interface, for parallel *(p)* and normal *(s)* polarization, respectively. The optical film thickness δ (i.e. the phase change experienced by the refracted beam, in traversing the film) follows from:

$$\delta = k_t d, \tag{7.4}$$

in which d is the film thickness and k_t the transverse propagation constant:

$$k_t = n_f k_o \cos \phi_f, \tag{7.5}$$

k_o is the vacuum wave number and ϕ_f the propagation angle of the refracted beam in the film:

$$\phi_f = \frac{1}{n_f} \arcsin \phi_o. \tag{7.6}$$

The terms $r_u^{p,s}$ and $r_l^{p,s}$ are the reflection coefficients at the upper (film-air) and the lower (substrate-film) interface for parallel *(p)* or normal *(s)* polarization, respectively:

$$r_{u,l}^{p} = \frac{n_b \cos \phi_a - n_a \cos \phi_b}{n_b \cos \phi_a + n_a \cos \phi_b}, \tag{7.7a}$$

$$r_{u,l}^{s} = \frac{n_a \cos \phi_a - n_b \cos \phi_b}{n_a \cos \phi_a + n_b \cos \phi_b}. \tag{7.7b}$$

A more detailed description is given by Azzam [1].

The parameters Δ and Ψ can be written as functions of the (complex) film and substrate refractive indices n_f and n_s, the film thickness *d*, and the incidence angle ϕ_o:

$$\Delta = \Delta(n_f, n_s, d, \phi_o), \tag{7.8a}$$
$$\Psi = \Psi(n_f, n_s, d, \phi_o). \tag{7.8b}$$

If n_s is known, and the film absorption (i.e. the imaginary part of n_f) can be neglected, n_f and d can be determined from a single measurement of Δ and Ψ using equations 7.8a and 7.8b. The influence of measurement errors in Δ and Ψ on the calculated values for n_f and d can be analyzed using these formulae, which are known as the Drude equations.

Accuracy can be improved by measuring Δ and Ψ at n incidence angles instead of at a single one. This yields $2n$ equations (the real and imaginary parts of equations 7.8a and 7.8b for each of the incidence angles) with two unknowns. The set is over-determined and, due to the unavoidable measurement errors, most probably incompatible. A Least-Squares Estimation (LSE) method is usually employed to find n_f and d from this set.

Loescher [3] and Humlícek [4] both analyzed the accuracy obtained using a LSE-method. This analysis assumes the measurement errors to be normally distributed and described by their covariance matrix. From this matrix the errors in the unknown parameters can be inferred, assuming equations 7.8a and 7.8b can be written in matrix form, i.e., can be linearly approximated around the observation point.

A more complicated approach is necessary to estimate the parameter errors in strongly non-linear regions, as will occur around the pseudo-Brewster angle (to be shown in the sequel). To avoid this problem we used a Bounded-Error (BE) approach, as introduced by Schweppe [13] and Witsenhausen [14], instead of a more conventional LSE method. The BE-approach has the additional advantage of permitting the validity of the assumptions made about the actually occurring measurement errors, which are often difficult to validate, to be easily tested. A non-linear BE-approach, as described by Smit and van Vliet [15,16], was employed here. A short description will be given below. A recent survey of the bounded-error estimation technique is presented by Walter and Piet-Lahanier [17].

The measurement result is represented as a hypercube B in a six-dimensional space, constructed from the six measurement entities (Δ and Ψ at *30°*, *50°* and *70°* incidence angles). The length of the edges is determined by the magnitude of the errors in the different entities. Because the possible values of Δ and Ψ are determined by only two parameters n_f and d, the model-response surface (defined by equations 7.8a and 7.8b) will be two-dimensional. Figure 7.2 illustrates the foregoing for a three-dimensional measurement space (a six-dimensional space being difficult to draw). The curved response surface represents the subspace of measurement values which are physically possible, the orthogonal lines indicating lines of constant n_f and d. The model membership set S, i.e. the region of values of n_f and d which could have caused the actual observation given the possible observation errors, is represented by the shaded area of the curved plane falling within the cube. To determine the size and shape of this membership set, we applied a boundary-detection algorithm as described by Freeman [18].

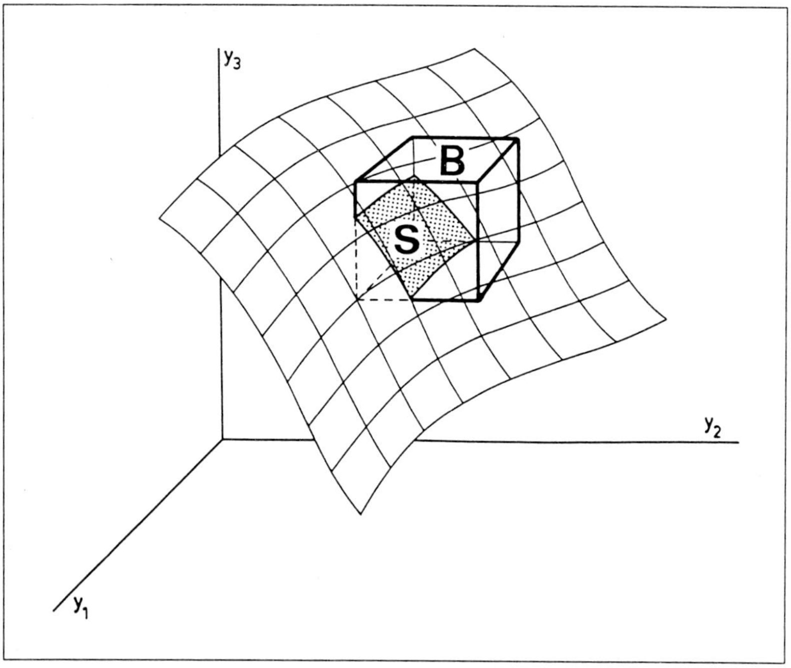

Figure 7.2 Cross-section (shaded region) of a two-dimensional model response surface with a three-dimensional uncertainty region.

Figure 7.3 shows an example of a membership set computed in this way. The maximal error in n_f and d can be read directly from such a plot. Further, the effect of a priori knowledge about one of the parameters, for example n_f, on the measurement accuracy of the other one can be immediately seen: the uncertainty region is reduced to the slice falling within the error bounds of n_f (the striped region).

If the measurement errors are estimated too optimistically, or if the description of equations 7.8a and 7.8b is not exact (e.g. due to film anisotropy or the presence of a thin film of adsorbed water) no intersection between the

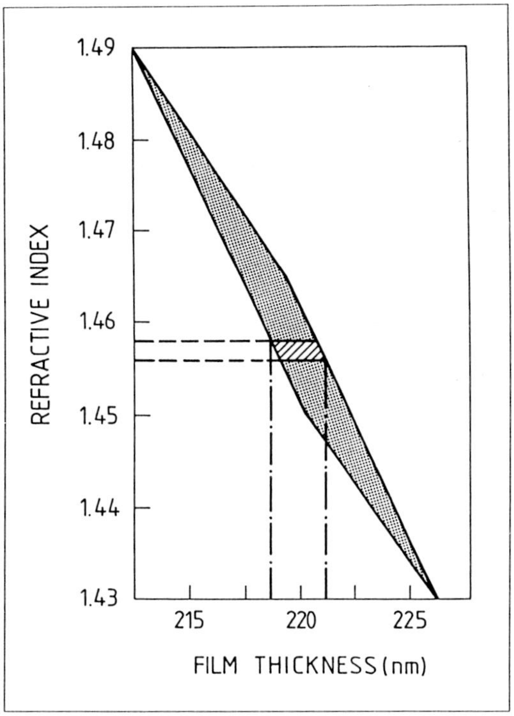

Figure 7.3 *Model membership set for n_f and d, with (striped) and without (dotted) a priori knowledge about the refractive index.*

cube B and the response plane may occur at all. This provides a means of testing the validity of our assumptions about the actual measurement accuracy for Δ and ψ. After computing the membership set S, as shown in figure 7.3, for a considerable number of independent observations, the smallest error-value of Δ and ψ for which all observations yield non-empty sets may be considered to be an indication of the actual measurement accuracy. This will be discussed in section 7.4.

7.3 Error propagation

To determine the error-propagation properties of the single- and three-angle measurements, we analyzed the effect of measurement errors in Δ and Ψ on the estimated values of n_f and d for a thermally oxidized SiO_2 layer on a silicon substrate. Conclusions will be extended to other film and substrate materials. All computations were performed using the program described by Smit and van Vliet [15,16].

7.3.1 Single-angle measurement

Figure 7.4 shows the magnitude of the maximum errors in the film thickness and the refractive index for thickness values ranging from 100 to 400 nm. The curves show the effect of an error of 1° both in Δ and Ψ. The first three curves represent the accuracy of results based on a single-angle measurement, for 30°, 50° and 70° incidence angles, respectively. From the figure it is clear that two situations have to be distinguished: for a film thickness between 110 and 140 nm the measurement errors are minimal, for the double thickness they are maximal. The first range corresponds to $\delta \simeq \pi/2$, i.e. the film thickness equals $\frac{1}{4}\lambda_t$ (λ_t being the transverse wavelength within the film, which is slightly angle dependent through equation 7.5). The second range corresponds to $\frac{1}{2}\lambda_t$ film thickness. These ranges will be referred to as the $\frac{1}{4}\lambda$ and $\frac{1}{2}\lambda$-range .

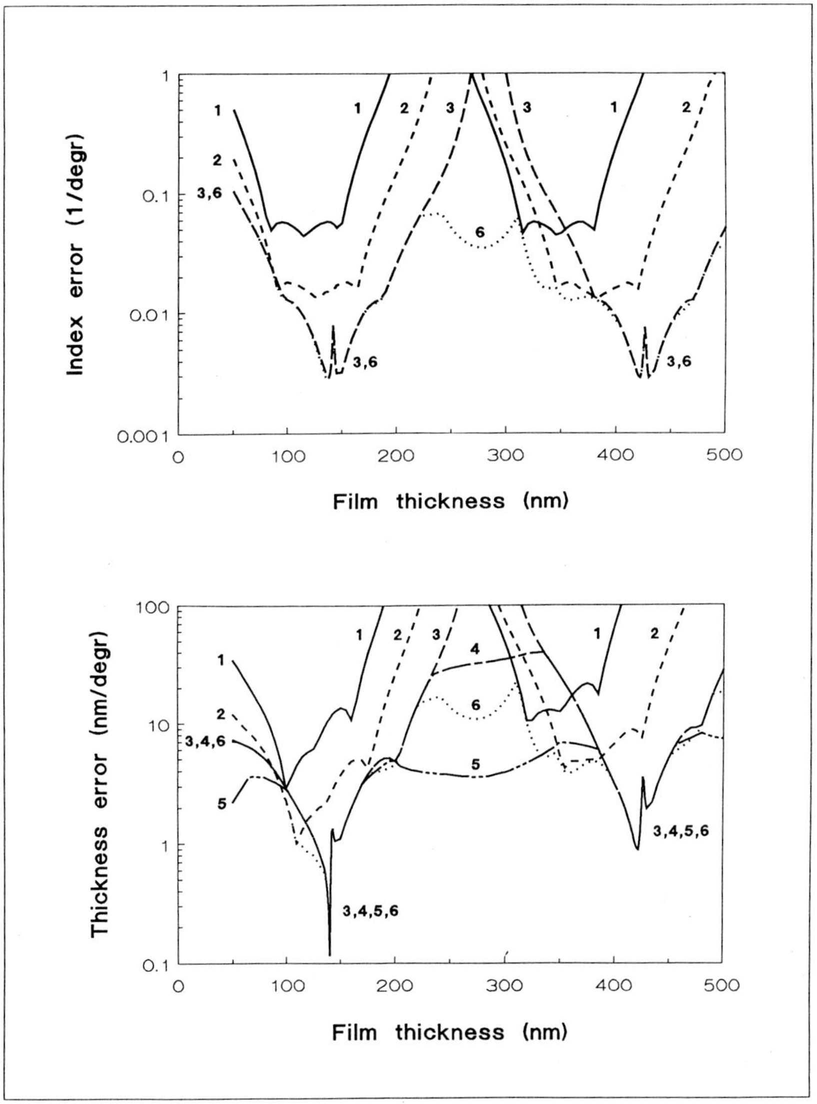

Figure 7.4 The measurement errors in (a) the refractive index and (b) the film thickness resulting from a measurement error of 1° in Δ and Ψ:

- *Curve 1: 30° incidence angle, no restriction on* n_f
- *Curve 2: 50° incidence angle, no restriction on* n_f
- *Curve 3: 70° incidence angle, no restriction on* n_f
- *Curve 4: 70° incidence angle,* $\delta n_f < 0.01$
- *Curve 5: 70° incidence angle,* $\delta n_f < 0.001$
- *Curve 6: 30°-50°-70° combined, no restriction on* n_f

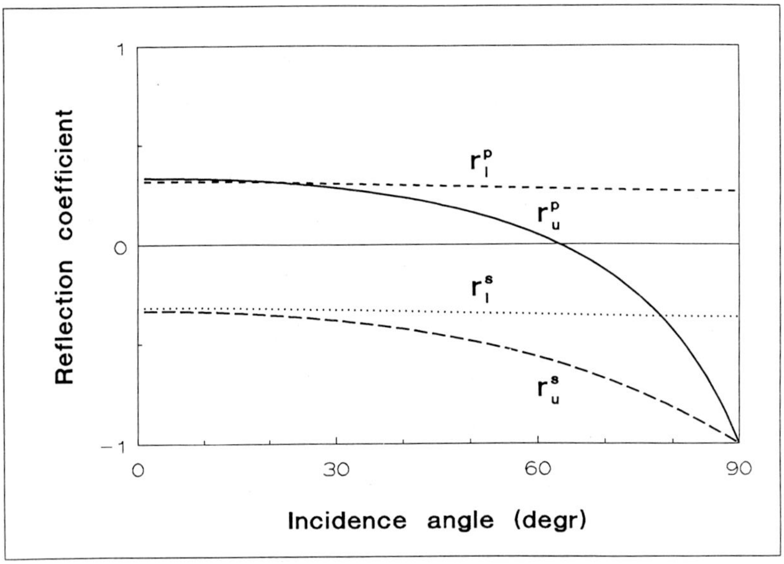

Figure 7.5 *The reflection coefficients at the upper (u) and the lower (l) interface of a ¼λ silicon nitride film (n≃2) on silicon substrate, both for parallel (p) and normal (s) polarization, measured outside the film.*

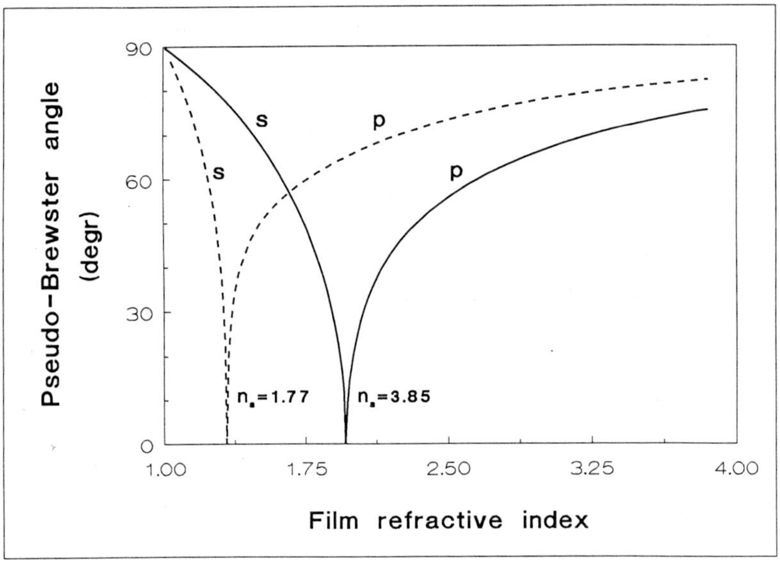

Figure 7.6 *The Pseudo Brewster angle as a function of the film index for a silicon and a sapphire substrate. The left-hand segment of the curves corresponds to the zero reflection occurring for parallel (s) polarization, the right-hand segment for normal (p) polarization.*

The conclusion that a film thickness of ½λ, or a multiple of ½λ, yields a maximal inaccuracy, whereas a film thickness of ¼λ (plus a multiple of ½λ) yields an optimal accuracy, applies to a wide variety of film-substrate combinations (see for example Humlícek [4]) and may be explained physically by the fact that the reflection of a low-loss ½λ-film does not differ from the reflection of the underlying substrate and thus contains no information about the film. Because the region around the point of optimal measurement accuracy is most interesting from a measurement point of view, the analysis may be restricted to ¼λ-films without great loss of generality.

Intuitively it is to be expected that accuracy will be optimal in parameter regions where ρ (equation 7.1) varies rapidly with n_f and d. This is the case at the so-called pseudo-Brewster angle of the film-covered substrate [1], i.e. the angle at which either ρ^p or ρ^s goes through zero. From equation 7.3 it is clear that for low-loss ¼λ films this angle occurs for $r_u^p \simeq r_l^p$ or $r_u^s \simeq r_l^s$.

Figure 7.5 illustrates the angle-dependence of these reflection coefficients for a silicon nitride film on silicon substrate. From the figure it can be seen that $r_u^p \simeq r_l^p$ at 20° incidence angle. For a film index $n_f \simeq \sqrt{n_s}$ the pseudo-Brewster angle occurs at normal incidence and ρ^p equals ρ^s for all incidence angles. This condition is optimal for anti-reflection coatings and corresponds to the ¼λ-transformator in microwave technique. The zero reflection condition occurs for ρ^s instead of ρ^p for lower values of n_f. Figure 7.6 shows the dependence of the pseudo Brewster angle on the film index for two different values of the substrate index (silicon and sapphire).

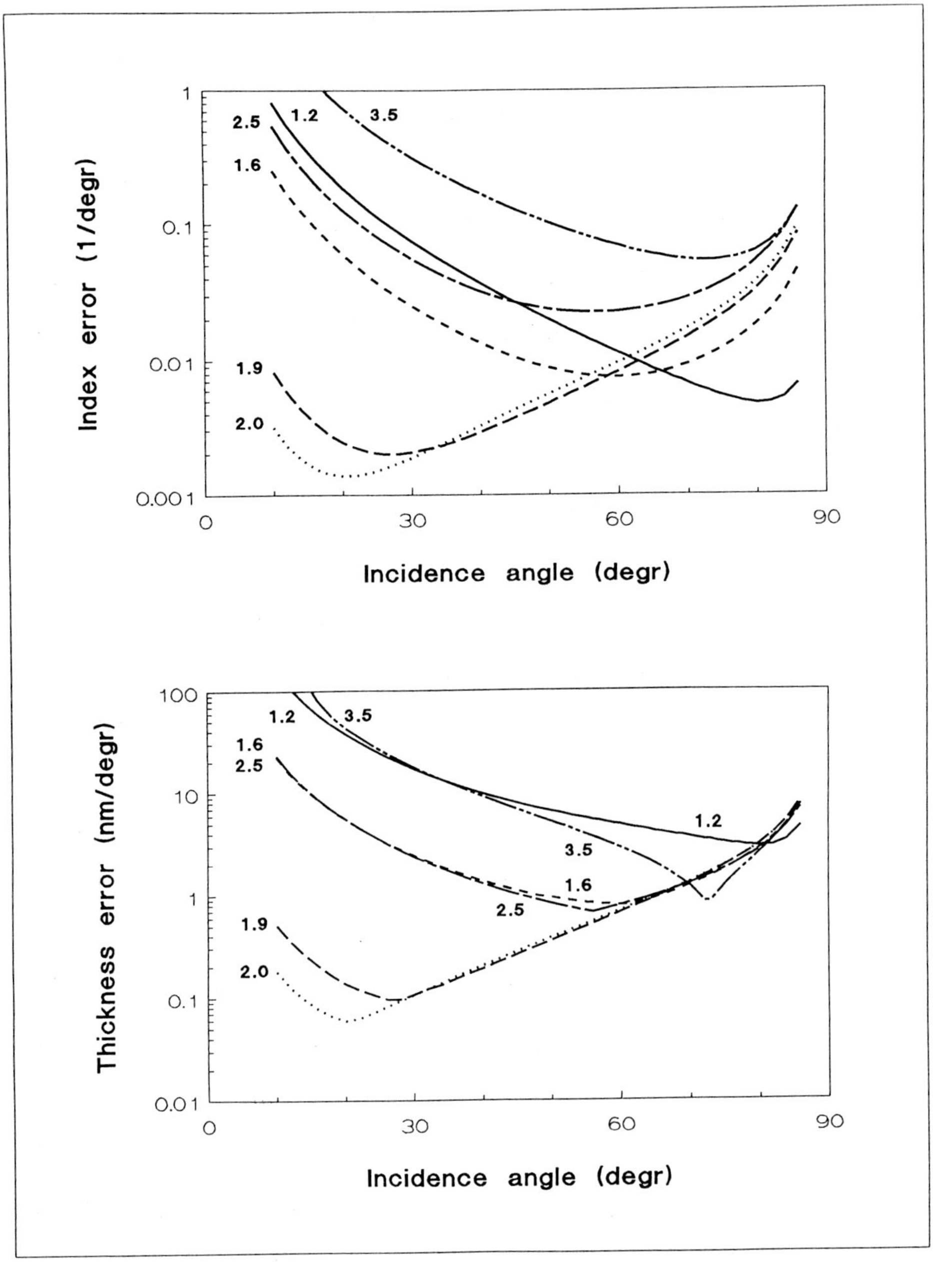

Figure 7.7 *The measurement errors in the film thickness (upper figure) and the refractive index (lower figure) for a film-thickness of ¼λ (i.e. at the accuracy optimum).*

To verify the relation between the pseudo-Brewster angle and the measurement accuracy, we computed the errors in n_f and d resulting from an error of 1° in Δ and Ψ for a series of film-indices. The results are shown in figure 7.7. Comparison of figures 7.6 and 7.7 confirms the intuitive expectation that measurement accuracy is optimal at this angle. From the figure it can be seen that the highest accuracy will be obtained at small pseudo-Brewster angles, i.e. for film indices close to $\sqrt{n_s}$.

7.3.2 Multi-angle measurement

The accuracy of a single-angle measurement collapses near ½λ film thickness. Because this condition occurs for only one specific incidence angle, measurement accuracy can be improved by utilizing the results obtained at a number of incidence angles simultaneously. The fourth curve in figure 7.4 presents the accuracy for the combined utilization of the measurement results obtained at 30°, 50°, and 70° incidence angles. For $d \simeq \frac{1}{4}\lambda$ the results are identical to the results obtained with a single 70° measurement. For $d \simeq \frac{1}{2}\lambda$, however, the accuracy improves by a factor of more than 10 compared to the best value obtained at one of the three angles. It is even better if compared only to the 70° measurement.

Addition of three other angles (40°, 60°, and 80°) yields only marginal improvement, which might be expected because the minima in figure 7.4 are rather broad, indicating a weak angle-dependence and, consequently, a high correlation between results obtained at closely spaced incidence angles. The choice of 30°, 50°, and 70° will, therefore, prove adequate for a wide range of film-substrate combinations.

Figure 7.8 shows the ½λ-points for three commonly employed materials. Construction of such diagrams for a variety materials is not difficult (using equation 7.5). The problematic film thicknesses for single-angle measurements can be easily read form the diagram. Measurement accuracy becomes less angle dependent when a combined three-angle measurement is employed, as can be seen from figure 7.4. Problems may be expected with regard to film thicknesses for which the ½λ-points for the various angles are close to each other. The accuracy of the combined method will severely degrade at these points, and additional angles may yield considerable improvements in accuracy. From the figure it can be seen , however, that this will not occur for film thicknesses under 1 μm.

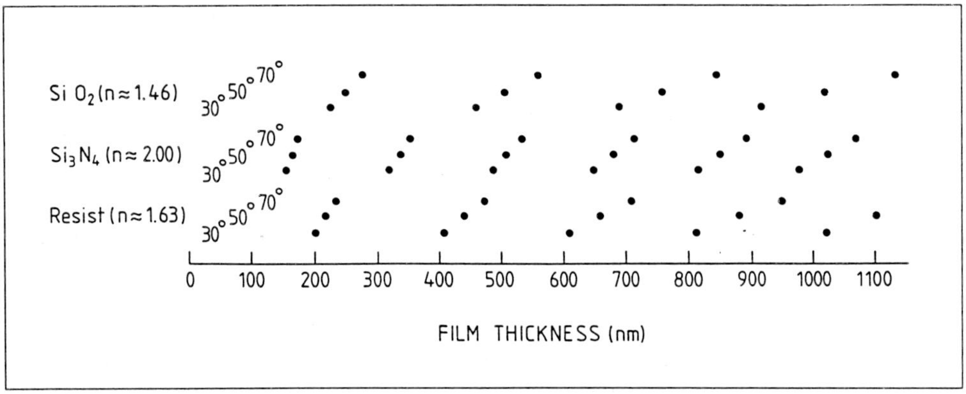

Figure 7.8 *½λ-Thicknesses for SiO_2, Si_3N_4 and photoresist (Hunt HPR-204), at 30°, 50° and 70° incidence angle.*

In conclusion, a multi-angle measurement method will yield a considerable improvement in accuracy relative to a single-angle measurement for film-thicknesses close to $\frac{1}{2}m\lambda$. It will be much less useful around $\frac{1}{4}\lambda+\frac{1}{2}m\lambda$. An additional advantage of the combined method is that the fundamental ambiguity of $\frac{1}{2}m\lambda_t$, which is inherent to the single-angle measurement, is strongly reduced in the combined approach due to the fact that λ_t is angle-dependent.

7.4 Accuracy analysis

To test the validity of the above described method, a series of measurements were made, each consisting of Δ and Ψ measured at 30°, 50°, and 70° incidence angles, on three silicon substrates covered with thermal oxide films of 145, 195, and 235 nm thickness, respectively (i.e. $0.25\lambda_t$, $0.35\lambda_t$, and $0.4\lambda_t$). The three substrates were obtained from one 2"-wafer by dicing it and etching the pieces to the required thickness in buffered HF. The $0.25\lambda_t$-substrate was measured at 9 different positions within 1 cm^2. The other substrates, which showed less spread in the measurement data, were measured at 6 positions. The number of measurement series thus totalled 21 three-angle series or 63 single-angle series.

Substrates were thoroughly cleaned before measurement (ultrasonic cleaning in a strongly alkaline soap solution, rinsing in demineralized water followed by spin-drying). The random variation of Δ and Ψ over the substrates was found to amount to ± 0.15° for Δ and ± 0.4° for Ψ. We tested the significance of these figures by substituting them for the maximum observation errors for all measured series. We found no solutions for n_f and d (i.e. all membership sets S were empty) for any of the 21 records. The error value had to be increased to 0.4° for both Δ and Ψ before the first nonempty set appeared. Increasing the error value to 0.6° caused all but three empty sets to disappear, which suggests that the systematic errors are considerably larger than the random ones. The remaining three empty sets all occurred with the 0.25λ-film; the last one disappeared on increasing the measurement error to 1.7°.

These results illustrate that in a practical measurement situation the errors may be considerably greater than those which are usually presented in the literature (0.01°-0.1°) for comparable measurement equipment. The discrepancy may be due to systematic errors occurring in the measurement of Δ and Ψ, but are more probably due to the deviation of the experimental measurement configuration from its idealized description in equations 7.1-7.8. This may occur due to adsorption of a thin water layer, depolarization due to surface scatter or birefringence due to tension in the film, to mention just a few possibilities.

Systematic errors will cause a displacement of the model response surface and the hypercube B, as shown in figure 7.1, relative to each other. If this displacement is directed along the response surface, the systematic error is not detectable with the present approach. If it is directed perpendicular to this surface, however, the lowest value of the observation errors for which none of the measurement series yields an empty membership set S, is indicative of the total effect of both random and systematic errors (a fixed ratio between the observation errors is assumed). By computing the membership set corresponding to the observation errors so-found, we obtain an estimate of the magnitude of parameter errors resulting from both systematic and random errors (except, of course, for the worst-case series on which the error estimate was based).

Assume that the number of measurements is much larger than the number of parameters. The projection on the response surface of the displacement induced by systematic errors is then likely to be small in relative terms. The error estimates produced by the above algorithm are, therefore, unlikely to be overly optimistic.

To investigate the practical value of the algorithm we proceeded as follows. We computed the refractive index value as the center of the uncertainty interval for all measurements. The error in this value was determined by comparing it to the value given in the literature for fused silica (1.4573, [19]) which is assumed to be close to the value for thermal oxide. In this way maximum errors were determined for both single and three-angle measurements for the three film thicknesses investigated.

Of a total of approximately 60 independent measurements (for the $0.25\lambda_t$ and the $0.35\lambda_t$-substrate the combined measurement and the 70°-measurement yield identical results, as might be expected from figure 7.4) we found only two cases in which the real measurement error, i.e. the difference between the literature value and the estimated value, exceeded the one predicted according to the above algorithm. In the worst of these, the real error was twice the one predicted. Both discrepancies occurred for the $0.25\lambda_t$ film, for measurement sets where the computed parameters coincided with the singularity observable in figure 7.4. If we exclude these measurements, the maximum errors found for the different measurement series were within 15-80% of the predicted errors, and for 60% of the series within 40-80%.

Although no general conclusions can be drawn from these experiments, the results indicate that in the field of ellipsometry the above algorithm may be applied with advantage to analyze the magnitude of both random and systematic errors.

The singularity occurring at the pseudo-Brewster angle for ¼λ-films is caused by the fact that if ρ^p or ρ^s crosses zero the ratio ρ and, consequently, the phase parameter Δ, experiences a phase jump of 180°. This corresponds to an extremely high sensitivity for the film-parameters resulting in the sharp dip observed in the 70° curve of figure 7.4. Because this extremely high sensitivity only occurs for values of ρ^p or ρ^s close to zero, the accuracy in measuring Δ will severely degrade, thus filling the accuracy dip. The problem might be circumvented by basing numerical methods on the real and imaginary portions of ρ, instead of on its amplitude and phase.

7.5 Conclusions

Accuracy in determining the parameters of a low-loss film on an absorbing substrate from measurement at a single incidence angle is very poor for any film thickness close to an integral number of half wavelengths (measured in the film along the normal) and optimal in between, independent of the incidence angle.

Accuracy at the optimum film-thickness (i.e. $\frac{1}{4}\lambda + \frac{1}{2}m\lambda$) is maximal for incidence angles close to the the pseudo-Brewster angle of the film-covered substrate. For silicon dioxide on silicon substrate and for epitaxially grown semiconductor films (high refractive index values for both film and substrate) the optimum incidence angle occurs between 70° and 75°. As the optimum range in the accuracy curves is broad, a set of 30°, 50°, and 70° provides optimal accuracy for a wide variety of film-substrate combinations.

For $\frac{1}{4}\lambda$-films addition of results measured at angles other than the optimal incidence angle (the pseudo Brewster angle) yields no improvement in accuracy. The accuracy loss near $\frac{1}{2}\lambda$ film thickness can be reduced considerably, however, by including results measured at other incidence angles having different transverse wavelengths λ_t within the film. Combined utilization of results measured at 30°, 50°, and 70° yields optimal results through a wide range of film parameters. The effect of adding more incidence angles is small.

Film parameters may be determined from the measurement results using either a Least-Squares Estimation method or a Bounded-Error Estimation method. The latter method also yields reliable accuracy information in parameter regions where the dependence of the observations on the parameters is highly nonlinear, as occurs for ¼λ-films near the pseudo-Brewster angle. Further, it provides a tool for analyzing the maximal measurement errors, both random and systematic, which occur in the applied measurement procedure.

References

7.1 R.M.A. Azzam and N.M. Bashara, Ellipsometry and polarized light (North-Holland, Amsterdam, 1977).

7.2 F.L. McCrackin, E.Passaglia, R.R. Stromberg, and H.L. Steinberg, "Measurement of the thickness and refractive index of very thin films and the optical properties of surfaces by ellipsometry", J. Res. Natl. Bur. Std., Vol. 67A, pp. 363-377, 1963.

7.3 D.H. Loescher, R.J. Detry, and M.J. Clauser, "Least-squares Analysis of the film-substrate problem in Ellipsometry", J. Opt. Soc. Am., Vol. 61, pp. 1230-1235, 1971.

7.4 J. Humlícek, "Sensitivity extrema in multiple-angle ellipsometry", J. Opt. Soc. Am. A, Vol. 2, pp. 713-722, 1985.

7.5 H.W. Dinges, "Determination of the optical properties of GaAs and InP by multiple angle of incidence ellipsometry", Journal de Physique, Coll. c10, Suppl. au no° 12, Tome 44, pp. 33-43, 1983.

7.6 D.E. Aspnes, J.B. Theeten, and R.P.H. Chang, "Nondestructive characterization of interface layers between Si or GaAs and their oxides by spectroscopic ellipsometry", J. Vac. Sci. Technol., Vol. 16, pp. 1374-13??, 1979.

7.7 B.Drévillon, J. Perrin, R. Marbot, A.Violet, and J.L. Dalby, "Fast polarization modulated ellipsometer using a microprocessor system for digital Fourier analysis", Rev. Sci. Instrum., Vol. 53, pp.969-977, 1982.

7.8 G.H. Bu-Abbud, N.M. Bashara and J.A. Woollam, "Variable wavelength, variable angle ellipsometry including a sensitivities correlation test", Thin Solid Films, Vol. 138, pp. 27-41, 1986.

7.9 R.G. Snyder, M.C. Rost, G.H. Bu-Abbud, J.A. Woollam and S.A. Alterovitz, "Variable angle of incidence spectroscopic ellipsometry: Application to GaAs-Al_xGa_{1-x}- As multiple heterostructures", J. Appl. Phys., Vol. 60, pp. 3293-3302, 1986.

7.10 J.W. Verhoof and M.K.Smit, "Error Analysis for Ellipsometry", Procs. 12th IMACS World Congress on Scientific Computation, July 18-22, 1988, Paris, Vol. 2, pp. 502-504.

7.11 M.K. Smit and J.W. Verhoof, "Accuracy analysis for multiple angle of incidence ellipsometry", Thin solid films,

7.12 M.K. Smit and J.W. Verhoof, "A bounded-error approach for accuracy analysis in ellipsometry", Mathematics and computers in simulation,

7.13 F.C. Schweppe, "Recursive state estimation: unknown but bounded errors and system inputs", IEEE Trans. Automatic Control, Vol. AC-13, pp. 22-28, 1968.

7.14 H.S. Witsenhausen, "Sets of possible states of linear systems given perturbed observations", IEEE Trans. Automatic Control, Vol. AC-13, pp. 556-558, 1968.

7.15 M.K. Smit, "A novel approach to the solution of indirect measurement problems with minimal error propagation", Measurement, Vol. 1 (4), pp. 181-190, 1983.

7.16 M.K. Smit and C.H. van Vliet, "OMTP: Fortran program for optimally solving indirect measurement problems", Measurement, Vol. 1 (4), pp. 209-211, 1983.

7.17 E. Walter and H. Piet-Lahanier, "Estimation of parameter bounds from bounded-error data: a survey", Procs. 12th IMACS World Congress on Scientific Computation, July 18-22, 1988, Paris, Vol. 2, pp. 467-472.

7.18 H. Freeman, "Boundary encoding and processing", in: B.S. Lipkin and A. Rosenfeld, Picture processing and psychopictorics (New York: Academic Press, 1970).

7.19 "American Institute of Physics Handbook", New York: McGraw Hill Book Company, 1982.

Chapter 8

Measurement technique
Prism coupling

The prism coupler is widely employed for coupling light into or out of planar waveguides for measurement purposes. In this chapter, a theoretical and experimental investigation into the coupling of light into or from narrow waveguides is described. Special attention is given to the effects of focusing. Further, methods and setups for measuring film properties, propagation constants, optical attenuation and transmission loss of films, waveguides and components are described.

8.1 Introduction

The methods most commonly used for coupling power into or out of a planar optical circuit are prism-coupling and butt-coupling. Prism-coupling was introduced for this purpose by Ulrich and Tien [1] and described in more detail by the same authors in later publications [2,3]. A compact description has been provided by Unger [4].

Prism-coupling has some important advantages over butt-coupling from an experimental point of view. It eliminates the alignment and field-matching problems which occur for butt coupling to narrow waveguides. It offers a potential for non-destructive measurement of optical attenuation, and it is suitable for simultaneous excitation of a series of waveguides, thus facilitating transmission measurements on optical components, as will be shown in the sequel.

Furthermore, prism-coupling is mode-selective and, consequently, also suitable for measurement of mode properties (propagation constant, mode profile), in multimode waveguides. This is well known for the case of transverse film modes, but much less so for the case of lateral waveguide modes. Excitation of lateral modes was reported by Millar et al. [5] and Simova et al. [6]. These reports lacked, however, quantitative details with respect to selectivity. A theoretical analysis of the coupling mechanism, including the excitation of lateral modes, was provided by Wang and Laybourn [7]. They calculated the coupling efficiency between a plane input wave with finite cross section and a number of waveguide modes as a function of both the incidence angle and the skew angle.

In this chapter their analysis will be extended to include the effects of diffraction, which will be shown to play a dominant role when light is coupled into narrow waveguides (section 3). The results described in this section were published earlier [8]. Section 2 recapitulates two-dimensional coupling theory. In the last section (section 4) the methods developed for measuring the properties of guided modes (propagation constant, attenuation, mode profile), waveguides (width and thickness, refractive index, refractive index contrast) and planar optical components (transmission coefficients, insertion loss) are discussed.

8.2 Transverse coupling mechanism

8.2.1 Coupling length

If a guided mode enters a region in which the top layer of the waveguide structure is covered with a prism (the region $z>0$ in figure 8.1), a small fraction of the guided flux in the exponential tails of the mode will strike the prism. If the prism index n_p is greater than the effective mode index N $(=\beta/k_o$, β is the propagation constant and k_o the vacuum wavenumber), this flux will radiate into the prism. The mode will thus become leaky and decay exponentially.

The length l_c over which the amplitude is reduced to its $1/e$ value is called the coupling length. It depends on the waveguide and mode parameters according to [4]:

$$l_c = exp(2gw/d)\,\frac{\beta d^2\,(t^2+w^2)\,(u^2+w^2)\,(1+\frac{1}{v}+\frac{1}{w})}{4t\,u^2\,w^2} \tag{1}$$

in which β is the propagation constant of the mode, d is the film thickness, g is the gap width, and u, v, w, and t are the normalized transverse propagation constants:

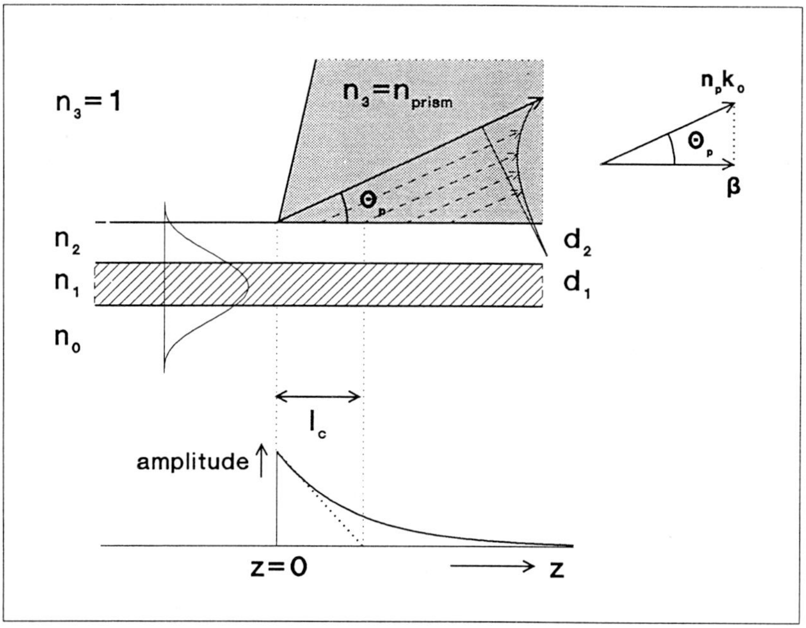

Figure 8.1 Transverse output coupling geometry.

$$u = k_o d\sqrt{(n_f^2 - N^2)}, \quad (1a)$$

$$v = k_o d\sqrt{(N^2 - n_s^2)}, \quad (1b)$$

$$w = k_o d\sqrt{(N^2 - n_g^2)}, \quad (1c)$$

$$t = k_o d\sqrt{(n_p^2 - N^2)}, \quad (1d)$$

In these formula's n_f, n_s, n_g, and n_p stand for the refractive indices of the waveguiding film, the substrate, the tunneling gap, and the prism respectively. Formula 8.1 is only valid for *TE*-polarization and is based on the assumption that the perturbation due to the presence of the prism is small, in which case the unperturbed value of the effective index *N* can be substituted. The latter assumption is justifiable for most practical cases.

For small contrasts the coupling length for *TM*-polarized modes will be close to the *TE*-value. More accurate estimates can be obtained by computing the complex effective index $N = N'\text{-}jN''$ for the prism-loaded waveguide structure. This is easily done using a transfer-matrix method as discussed in chapter 3. The coupling length l_c then follows as:

$$l_c = \frac{1}{\alpha} = \frac{1}{N'' k_o}, \qquad (2)$$

in which α is the attenuation coefficient.

Formula 8.2 is also useful for computing the prism coupling length for three-dimensional waveguide modes in SiO_2-covered ridge guides. If the effective index method, as discussed in chapter 3, is applied to the complex effective indices which are found for the prism-covered regions, then the coupling length can be derived directly from the imaginary part of the effective index computed for the ridge guide mode.

8.2.2 Coupling angle

Light leaves the waveguide under an angle θ_o for which $n_p k_o \cos\theta_o = \beta$ (phase match) according to the continuity conditions for the tangential field components in the plane $x=0$, so that:

$$\theta_o = \arccos \frac{N}{n_p}. \qquad (3)$$

From this formula it is apparent that only modes with effective indices $N<n_p$ can be coupled into or out of the waveguide.

8.2.3 Coupling efficiency

Because radiation is proportional to power level in the waveguide, the output beam will have an exponential profile as depicted in figure 8.1. In the absence of absorption, 100% of the guided power will be coupled to the output beam. For reasons of reciprocity, an input beam with the mirrored exponential profile and an incidence angle θ_o, as shown in figure 8.2, will be coupled to a guided mode with 100% efficiency if the exponential beam is properly aligned relative to the prism corner. Such a beam will be described by its

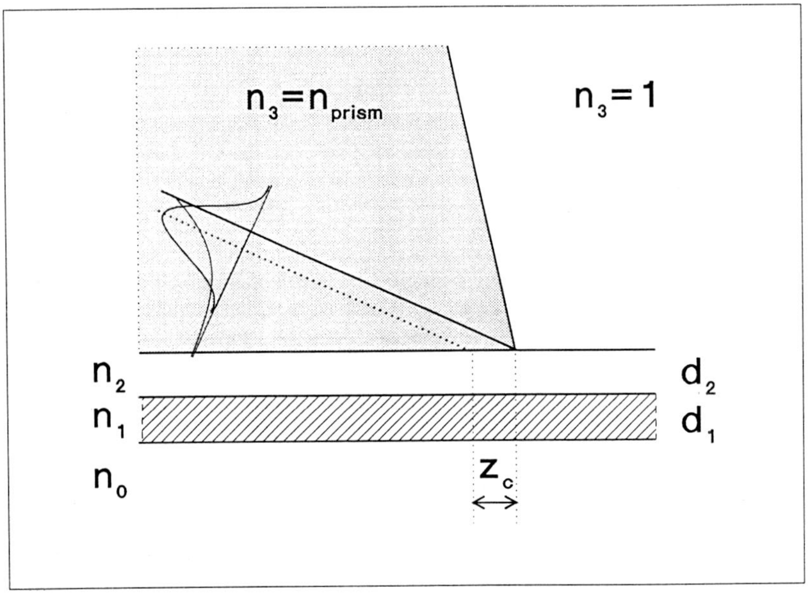

Figure 8.2 Transverse input coupling geometry.

(complex) z-dependence $U_e(z)$ in the plane $x=0$. If a differently shaped input beam $U_i(z)$ is employed, the coupling efficiency is determined by the overlap of the two beams, as computed using the well-known coupling integral:

$$\eta = \frac{|\int U_i(z)\; U_e^*(z)\; dz|^2}{\int |U_i(z)|^2 dz \;\; \int |U_e(z)|^2 dz}, \tag{4}$$

in which the integration is over the y-z plane at the bottom of the prism. With a Gaussian input beam $|U_i(z)| = exp(-z^2/w_z^2)$ a maximum coupling efficiency of about 80% is obtained if [4]:

$$l_c = 1.48\, w_z, \tag{5a}$$

$$z_c = 0.73\, w_z, \tag{5b}$$

z_c being the distance from the beam center to the prism corner. For the 0.25 μm thick waveguide films applied in our experiments, a silicon-dioxide top-layer thickness $g = 0.57\ \mu m$ yields a coupling length $l_c \simeq 1.5\ mm$, which coordinates well with the width $w_z \simeq 1\ mm$ of the applied input beam.

8.2.4 Coupling selectivity

Coupling selectivity is derived from equation 8.4 by considering the θ-dependence of η. It should be noted that the maximum coupling efficiency occurs for the value of θ for which the periodicity $exp(-jn_p k_o z cos\theta)$ of $U_i(z)$ just cancels the periodicity $exp(+j\beta z)$ of $U_e^*(z)$ in the integrand of the numerator. This means that a small incidence angle deviation $\Delta\theta$ will cause a term

$$exp\{-j \left.\frac{\partial(n_p k_o \cos\theta)}{\partial\theta}\right|_{\theta=\theta_o} \Delta\theta\, z\} = exp\{jn_p k_o sin(\theta_o) z\Delta\theta\}, \qquad (6)$$

to appear in the integrand. This exponential term has no effect on the denominator, so that η follows as:

$$\eta(\Delta\theta) = \eta(0) \frac{|\int p(z) exp\{j n_p k_o sin(\theta_o) z \Delta\theta\} z|^2}{\int | p(z) |^2 dz}, \qquad (7)$$

where

$$p(z) = U_i(z) U_e^*(z). \qquad (8)$$

The integral in the numerator is a Fourier integral, so that $\eta(\Delta\theta)$ follows as the square of the Fourier transform $P\{n_p k_o sin(\theta_o)\Delta\theta\}$ of $p(z)$. To obtain a simple expression for p(z) we approximate it as if U_e and U_i were both Gaussian:

$$p(z) = exp(-z^2/w_p^2), \qquad w_p = (1/w_e^2 + 1/w_i^2)^{½}. \qquad (9)$$

Employing this approximation, $\eta(\Delta\theta)$ follows as:

$$\eta(\Delta\theta) = \eta(0)\, [exp\{-n_p^2 k_o^2 w_p^2 sin^2(\theta_o)(\Delta\theta)^2/4\}]^2. \qquad (10)$$

From this formula the effective (Gaussian) width $w_{\theta i}$ of the coupling efficiency, which determines the coupling selectivity, as measured ***inside*** the prism, is approximated as

$$w_{\theta i} = \frac{2}{n_p k_o \sin\theta_o w_p} . \tag{11}$$

Substitution of $w_e = l_c/2$ and $w_i = w_z$ into equation 8.10 yields:

$$w_{\theta i} \simeq \frac{2(1/w_z^2 + 4/l_c^2)^{1/2}}{n_p k_o \sin\theta_o} . \tag{12a}$$

The approximate angular width $w_{\theta e}$ ***outside*** the prism follows by multiplication with n_p:

$$w_{\theta e} \simeq \frac{2(1/w_z^2 + 4/l_c^2)^{1/2}}{k_o \sin\theta_o} . \tag{12b}$$

To verify the validity of the above formula in describing the coupling selectivity we measured the width of the coupling curve at the $1/e^2$ points for three aluminum oxide films with thicknesses 0.16, 0.205, and 0.25 μm, which were covered with a 0.6 μm SiO_2 film. The experimental results are presented in figure 8.3. The agreement with the predicted curve, according to eq. 8.12b, which is also shown in the figure, is quite satisfactory.

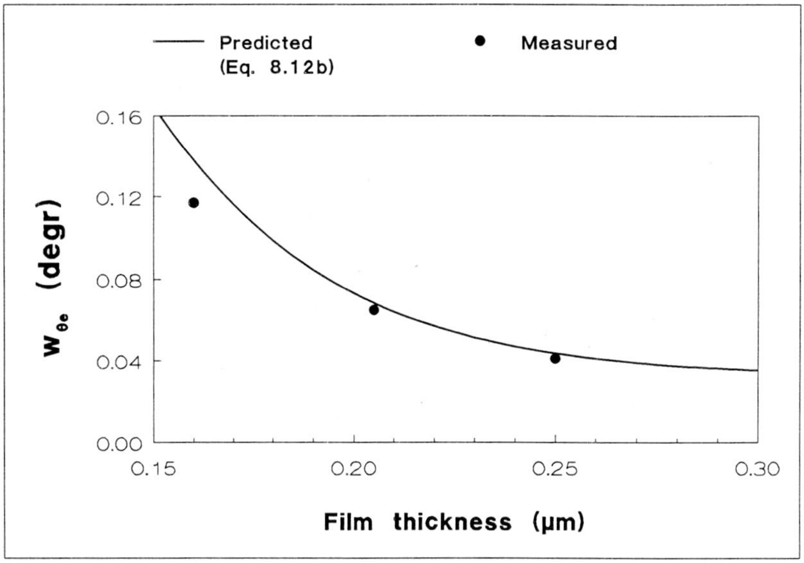

Figure 8.3 *Angular width* $w_{\theta e}$ *of the coupling curve as a function of the film thickness d* (g=0.6 μm, $n_s = n_c$=1.457, n_f=1.690, λ=0.6328 μm).

8.3 Three-dimensional coupling mechanism

8.3.1 Coupling efficiency

In the three-dimensional situation, the input beam at the bottom of the prism has to be matched to the lateral profile of the exponentially increasing mode. This provides 100% coupling efficiency. At first glance it might be expected that an optimal coupling efficiency would be obtained if the waist of the input beam matches the lateral width of the guided mode. This appears not to be true, however, as will be explained below.

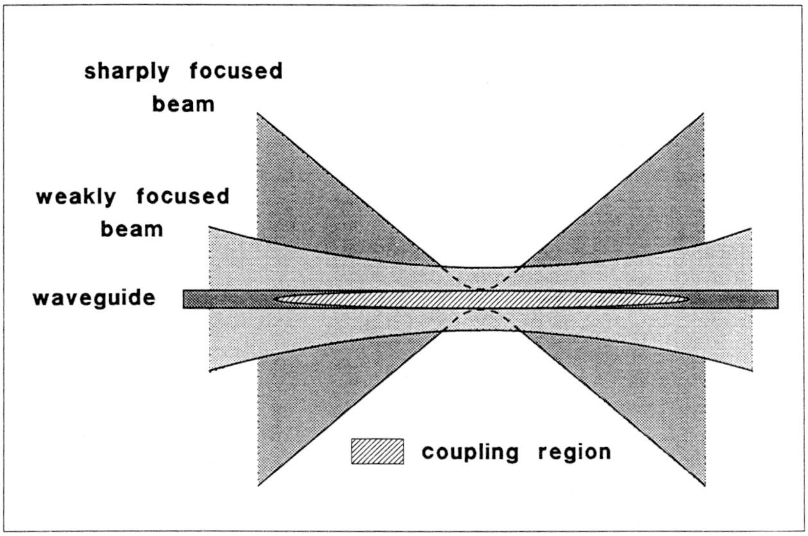

Figure 8.4 *Projection of a sharply and a weakly focused input beam onto the substrate surface.*

Figure 8.4 shows the projection of a sharply and a weakly focused (gaussian) input beam onto the bottom of the prism. Due to diffraction it is not possible to match its lateral profile to the lateral mode profile through the whole coupling region. If a sharply focused beam is employed the field match will rapidly degrade away from the beam waist. If a weakly focused beam is employed the field match in the waist will be reduced, but it will also degrade less rapidly. To determine the configuration for optimum coupling the optimal overlap of the two fields in the *y-z* plane has to be determined as a function of the input beam waist (i.e. the focusing angle).

Coupling efficiency for the three-dimensional coupling situation is inferred in much the same way as for the two-dimensional case by applying the overlap-integral theorem in two dimensions:

$$\eta = \frac{\left| \int\int U_i^*(y,z)\, U_e(y,z)\, dydz \right|^2}{\int\int \left| U_i(y,z) \right|^2 dydz \;\; \int\int \left| U_e(y,z) \right|^2 dydz} \, . \qquad (13)$$

In this formula $U_i\,(y,z)$ describes the *y-z* dependence of the input beam at the bottom of the prism (see figure 8.4), which is found by substituting $x=0$ into the description of the incident three-dimensional Gaussian beam. It should be noted that the beam-parameters for the two transverse dimensions differ because of the lateral focusing. $U_e(y,z)$ represents the field required for 100% coupling efficiency. Its (exponential) *z*-dependence follows from the coupling length of the prism-loaded waveguide. To obtain a simple expression for the overlap integral we applied a Gaussian approximation to the lateral mode profile, as described in chapter 3, section 3.2.1B.

In the three-dimensional situation η is dependent on the incidence angle as well as on the skew angle of the input beam. In our analysis we assumed the skew-angle to be zero, which is the optimal situation for exciting the fundamental mode. For exciting odd lateral modes a (small) skew angle is essential; experiments show that the effects on selectivity are small, however.

The evaluation of coupling efficiency according to equation 8.13 is straightforward, but slightly laborious. An integral expression for the coupling efficiency is derived in the appendix at the end of this chapter. Numerical evaluation of this expression as a function of the lateral input beam waist w_{yo} with the effective mode width w_m as a parameter yields the results plotted in figure 8.5 with solid lines. We substituted the optimal values for the two-dimensional coupling case (equation 8.5) for the coupling length l_c and the optimal offset z_c of the input beam center relative to the corner of the prism.

The following conclusions can be drawn from figure 8.5. As might be expected, for very wide waveguides the maximum coupling efficiency approaches 80% as in the two-dimensional situation. The maximum occurs at

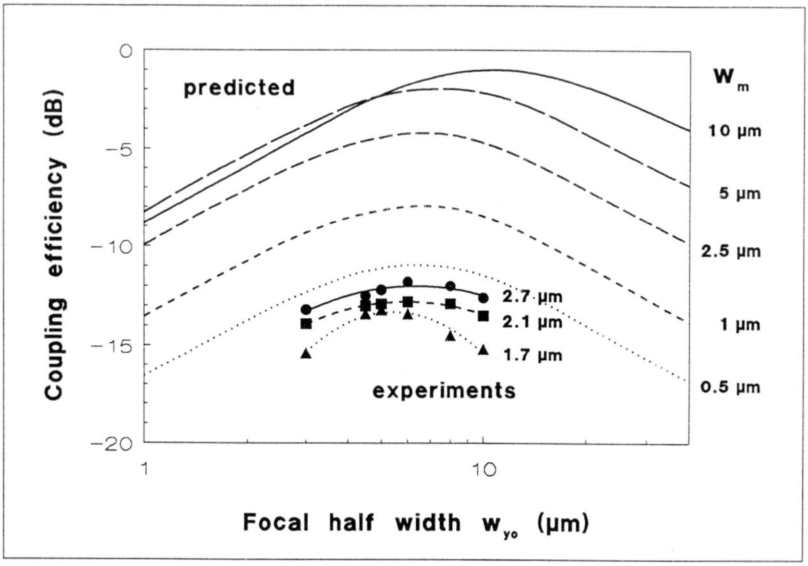

Figure 8.5 The maximum coupling efficiency as a function of the focal width w_{yo}*, for various values of the effective mode width* w_m*. Solid curves are computed. Dots represent experimental results.*

$w_{yo} = w_m$, i.e. the input beam and the mode have the same width. For smaller mode widths the maximum coupling efficiency is considerably reduced and the position of the maximum is fixed approximately at $w_{yo} \simeq 6\ \mu m$. As can be seen, it makes no sense to apply focal (half) widths less than 6 μm for narrow waveguides (for $\lambda = 633\ nm$). Furthermore, the efficiency loss in employing considerably wider beams (20-30 μm) is relatively small.

Computed results were experimentally checked for a number of focal widths and waveguide widths. The experimentally found values are also indicated in figure 8.5. The shape of the experimental curve and the position of the maximum compare well with the theoretical results. The absolute levels differ a few decibels; we did not invest much time in absolute calibration.

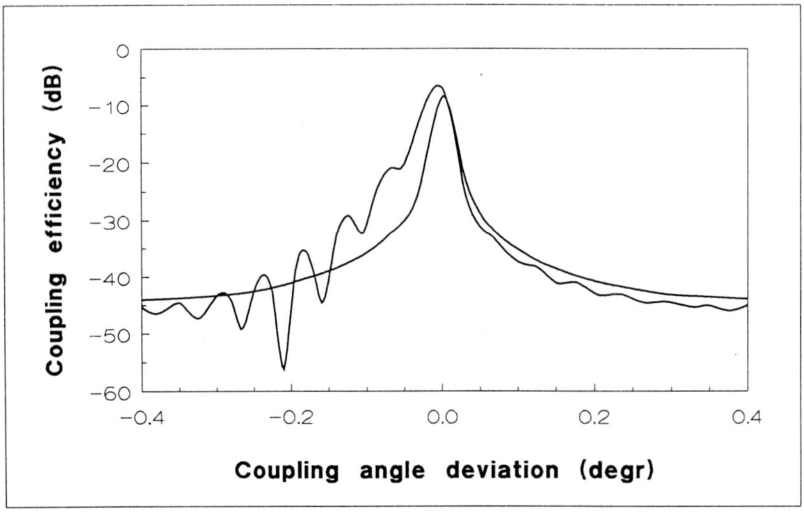

Figure 8.6 Coupling efficiency as a function of the angular deviation $\Delta\theta$.

8.3.2 Coupling selectivity

Focussing also affects the selectivity of the coupling. Just as in the two-dimensional case, a small incidence angle deviation $\Delta\theta$ contributes a factor $exp(+jn_p k_o z\Delta\theta sin\theta)$ to the coupling integral (equation 8.13). Figure 8.6 shows the angle dependence of η for two different values of the ratio between the input beam waist w_{yo} and the (Gaussian) modal half width w_m (for $w_m \simeq 1.7$ μm). For a wide input beam the width is approximately the same as for the transverse coupling configuration. This might be expected because of the planar character of the input beam. The angular width of the coupling curve increases, though not dramatically, for beam widths comparable to the mode width in the focal plane

These results were verified experimentally on a 3.5 μm wide waveguide with $w_m \simeq 1.7\ \mu m$. Figure 8.7 (upper graph) shows the measured coupling spectrum of the first three lateral modes of the waveguide excited with a wide beam $(w_{yo}/w_m \simeq 12)$. The half width of the TE_{00}^{l} curve at the $1/e^2$ level is approximately 0.00027 rad, corresponding to $w_\eta \simeq 0.02°$. Figure 8.6 also predicts 0.02°. The lower graph shown in figure 8.7 depicts the spectrum resulting from excitation with a sharply focused beam $(w_{yo}/w_m \simeq 1.5)$. The broadening of the spectrum $(w_\eta \simeq 0.07°)$ appears to be considerably greater than predicted by figure 8.6 $(w_\eta \simeq 0.03°)$. This discrepancy may be explained as follows.

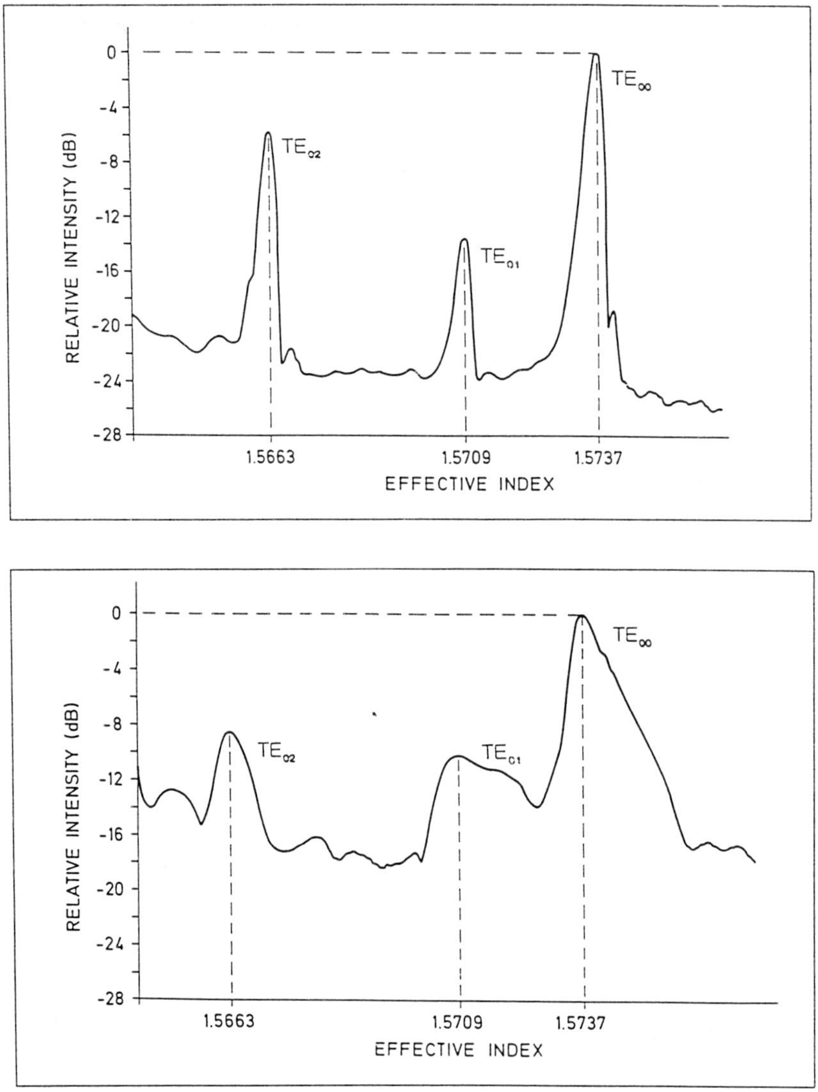

Figure 8.7 *Experimental record of the coupling efficiency for three lateral modes as a function of* θ *for two focal widths (and* $w_m \simeq 1.7\ \mu m$*):*

a) : $w_{yo}/w_m \simeq 12$ *(upper figure)*
b) : $w_{yo}/w_m \simeq 1.5$ *(lower figure)*

Wang and Laybourn [7] have calculated and Miller at al. [5], Simova et al. [6], and the present authors have demonstrated experimentally on the basis of the diffraction pattern of the field coupled out of a waveguide that radiation does not occur along a plane at an angle θ with the bottom of the prism, but along a conical surface. This can be understood by considering the spectrum of plane waves originating from the coupling region. The magnitude of the wave vectors is the same for all waves ($|k|=n_p k_o$). The same holds for their *z*-components, which equal β (phase-match condition). Consequently, the transverse component will decrease with increasing magnitude of the lateral one. This means that the outgoing radiation will be curved along a conical surface with the apex in the coupling region.

Taking reciprocity into account it may therefore be concluded that, in order to obtain good efficiency, the input beam should be focused along the same conical surface. This is difficult to realize in practice. If we do not fulfill this requirement, the field-match in the coupling region will be violated. This is probably the origin of the spectral broadening observed in the lower part of figure 8.6.

The direction of the broadening is in agreement with this explanation. The curvature of the conical surface of the output field is in the direction of smaller values of θ, i.e. higher values of N. This means, conversely, that if we employ a non-curved input beam, the broadening of the coupling spectrum will occur in the direction of higher N as has been observed.

8.3.3 Conclusions

Prism coupling of light into narrow optical waveguides is only possible if high coupling losses due to diffraction effects in the coupling region can be accepted. A wide waveguide should be applied in the coupling region and afterwards tapered to the required width if high coupling efficiencies are required. The application of an "oversized" input beam is advantageous, a small focal width reducing both selectivity and coupling efficiency if a maximal coupling efficiency is not required. In addition, a large focal width reduces the sensitivity of the coupling to environmental vibrations.

8.4 Measurement methods

8.4.1 Propagation constant and effective refractive index

The propagation constant can be determined from the coupling angle through equation 8.3: $\beta=n_p k_o \cos\theta_o$, in which θ_o is the angle between the input beam and the substrate (measured inside the prism) for which coupling is maximal. Accounting for the refraction at the prism front face, we find

$$N = \beta/k_o = n_p \cos\left[\frac{1}{n_p} \arcsin\{\theta_e - (\pi/2-\alpha_p)\} + (\pi/2-\alpha_p)\right], \qquad (14)$$

in which θ_e is the externally measured coupling angle, and α_p is the angle between the front face and the base of the prism (see figure 8.2).

Figure 8.8 shows the setup for measuring propagation constants. The laser signal is lead through a ¼λ-plate, a chopper, a beam splitter, and a cylindrical lens to the input prism. The circular polarization state behind the ¼λ-plate allows for excitation of both *TE*- and *TM*-polarized modes. The output prism is provided with a photodiode. The photocurrent is applied to a lock-in amplifier, which is locked on the chopper reference signal. The measurement accuracy can be enhanced by dividing the detected signal by the reference signal behind the beam splitter, thus eliminating the influence of source level fluctuations.

Both prisms are made of SF13 glass ($n=1.7356$ @ $633\ nm$), and coated with a chromium film on opposite faces in order to avoid direct transmission of scattered light. Prism base angles should be chosen such that multiple reflections inside the prism cannot hit the prism base under the same angle as the direct beam. We use a front face angle of 60° and a back face angle of 75°. Prism faces which are not used for transmission of light are made diffuse in order to further reduce the risk of coherent interference with multiply reflected beams.

To avoid the reproducibility problems and the poor local coupling uniformity which are inherent to the use of an air gap as tunneling layer, the silicon-dioxide top layer is used as a tunneling gap instead. Methylene diodide (CH_2I_2), which has a refractive index ($n \simeq 1.74$) close to the prism index, is applied under the prism in order to prevent the occurrence of an additional air gap between the silicon dioxide top layer and the prism.

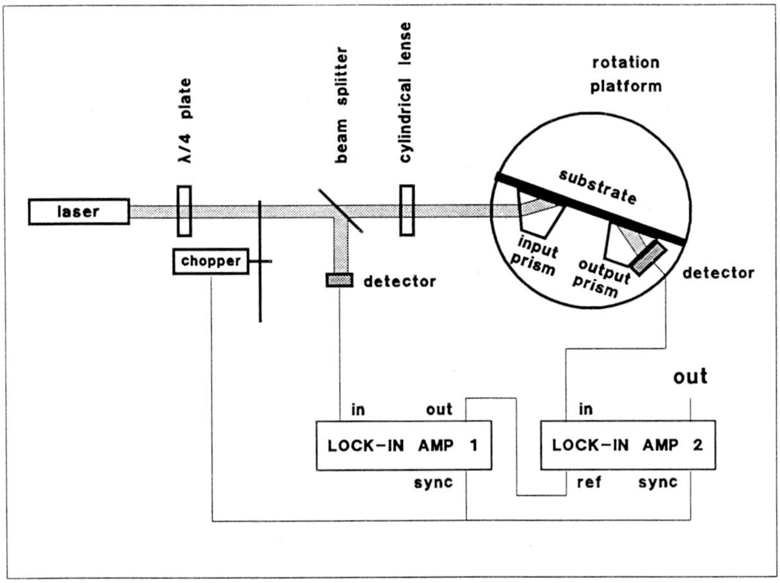

Figure 8.8 Schematic top view of the prism-coupling setup. The photograph shows a close-up of the substrate holder with the coupling prisms.

Coupling uniformity obtained in this way is in the order of 0.1 dB over several mm's. The latter uniformity is only obtained if the coupling angle is continuously adjusted to maximum coupling efficiency. Without adjustment a local variation of the effective index in the order of 10^{-3} will cause a reduction of the coupling efficiency by several decibels.

The substrate and both prisms are mounted on a rotation stage with a measurement precision of ±1'. Calibration is performed by positioning the substrate perpendicular to the input beam and aligning the incident and the reflected beam, and introduces another ±1' measurement error. The maximum error in measuring the coupling angle is thus estimated to be within ±2'. With $n_p = 1.74$ and $N \simeq 1.6$ the corresponding absolute accuracy in measuring the effective index N is $\pm 4.10^{-4}$. Comparably, the variation in the effective index of a 0.25 µm thick film, sputtered with the standard process, over 1 cm² is approximately $\pm 2.10^{-3}$; the fabrication tolerance has the same value.

The measurement resolution is determined by the angular width of the coupling curve. For an optimal coupling length $l_c \simeq 1.5\ mm$ the effective angular half width $w_{\theta e}$ (eq. 8.12b) is 10^{-3}. The detection resolution is approximately 0.1 dB. If the coupling curve is assumed to be quadratic, the full width at the - 0.1 dB level $\simeq 2.10^{-4}$, which is a limit for the measurement resolution. This value is in agreement with the results of reproducibility experiments carried out by a number of researchers. An effective-index variation of 2.10^{-4} corresponds to a thickness variation of 0.5 nm in the standard film thickness of 0.25 µm., i.e. in the order of molecular dimensions!

The ultimate measurement accuracy being limited by the measurement resolution ($\Delta N=\pm 2.10^{-4}$), it is concluded that the absolute accuracy obtained with the present setup ($\Delta N=\pm 4.10^{-4}$) is close enough to the ultimate value, and sufficient when compared to fabrication tolerance and variation over 1 cm² (both $\Delta N=\pm 2.10^{-3}$).

8.4.2 Waveguide parameters

The thickness and the refractive index of a single film on a silicon substrate are easily determined with an ellipsometer. Determination of the film parameters is more complicated for a three-layer structure. If the refractive indices of the silicon dioxide cladding layers are known, the parameters of the aluminum oxide film can be determined easily and non-destructively from the effective indices measured for two different modes (usually TE_0 and TM_0).

If the refractive indices n_s and n_c of the substrate and cover layer are considered to be constants, the dispersion relation (equation 3.2) can be written as:

$$V = f(N, n_f, m, q), \qquad (15)$$

in which V is the V-parameter, n_f the film index, N the effective index, m the order of the mode and q the polarization parameter as described in chapter 3. V being independent of mode number and type, the following relation holds for each two guided modes:

$$\Delta V = f(N_1, n_f, m_1, q_1) - f(N_2, n_f, m_2, q_2) = 0, \tag{16}$$

in which $q_{1,2}$, $N_{1,2}$, and $m_{1,2}$ stand for the polarization parameter ($q=0$ for *TE* and $q=2$ for *TM*-polarization), the measured effective index, and the order of the two modes. From this relation n_f can be solved using a numerical root-finding algorithm, a variety of which are available in most numerical libraries. Once n_f is found, V follows from eq. 8.15, and the film thickness d is readily computed from:

$$d = V/\{k_o \sqrt{(n_f^2 - n_s^2)}\}. \tag{17}$$

The above procedure is based on the assumption that all layers have a uniform isotropic index profile. To test this assumption measurements were performed on a number of layers of different thickness, for which the method is expected to yield the same refractive indices. The refractive indices computed from the TE_0 and the TM_0 mode increase from 1.68 for a 125 nm film to 1.69 for a 250 nm film. For thicker films the index decreases again and the corresponding predicted film thickness assumes unrealistic values. This suggests that our films are not uniform or isotropic.

To investigate the birefringence of our films we determined the parameters of a thick (bimodal) film both from the TE_0 and the TE_1-modes, which gives us the *TE*-polarized refractive index, as well as from the TM_0 and the TM_1-modes. This leads to an average refractive index of 1.701 and a birefringence

n_{TE}-$n_{TM} \simeq -2.10^{-3}$ (after annealing). The average value agrees well with ellipsometrically measured values, which, however, are based on the assumption of isotropy. The thickness, computed after substitution of the above index values, is far more proportional to the sputtering time than thickness values computed on the assumption of isotropy. This supports the birefringent hypothesis. Measurement of the effective indices of *TE*- and *TM*-modes on both sides of an etched step, with the assumption that the optical properties are equal on both sides of the step, leads to different values, however, thus indicating that the birefringent model is too simple. Further refinement may require the introduction of a non-uniform index profile or birefringent cladding layers.

Because the analysis and the design of planar components is based on effective indices which are directly measured, rather than the optical and geometrical film parameters, it was not considered useful to invest too much experimental work in further refinement of the description of the waveguide structure.

8.4.3 Optical attenuation

Optical attenuation of both ridge guides and slab waveguides can be determined by coupling light into the waveguide with a fixed input prism and recording the intensity along the waveguide with a sliding output prism. The matching liquid (CH_2I_2) applied provides good optical contact between waveguide and prism during sliding without causing mechanical damage. The output prism is moved using an electronic translator. The detected output power is recorded logarithmically against the translation.

Figure 8.9 shows an experimental record demonstrating the quality of the measurement procedure. The measurement accuracy is dependent on the prism-coupling length. The reproducibility of the input coupling degrades severely for short coupling lengths. For long coupling lengths the output prism detects only part of the total power, which also reduces stability. For coupling lengths between 1 and 2 mm and a measurement range of 10 mm the measurement accuracy is in the order of 0.1 dB/cm.

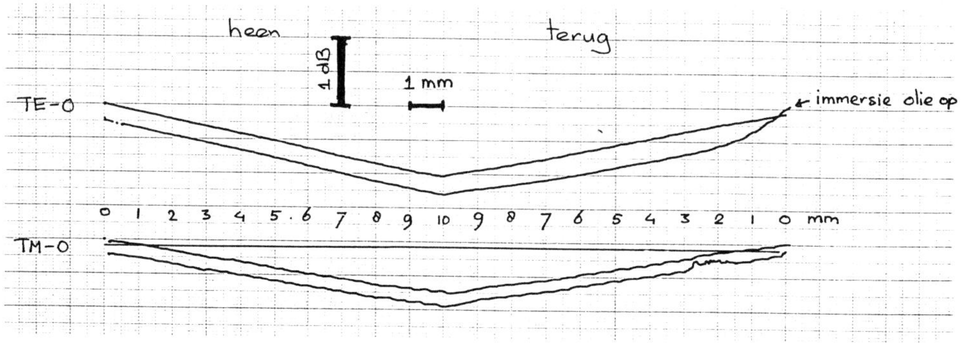

Figure 8.9 Example of a (logarithmic) attenuation recording. The coarse grid corresponds to 0.5 dB/division vertically, and 1 mm/div horizontally.

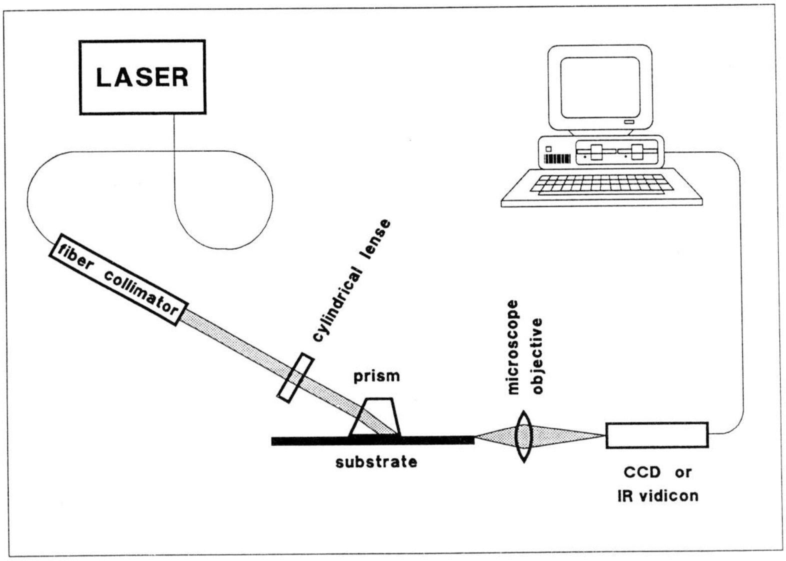

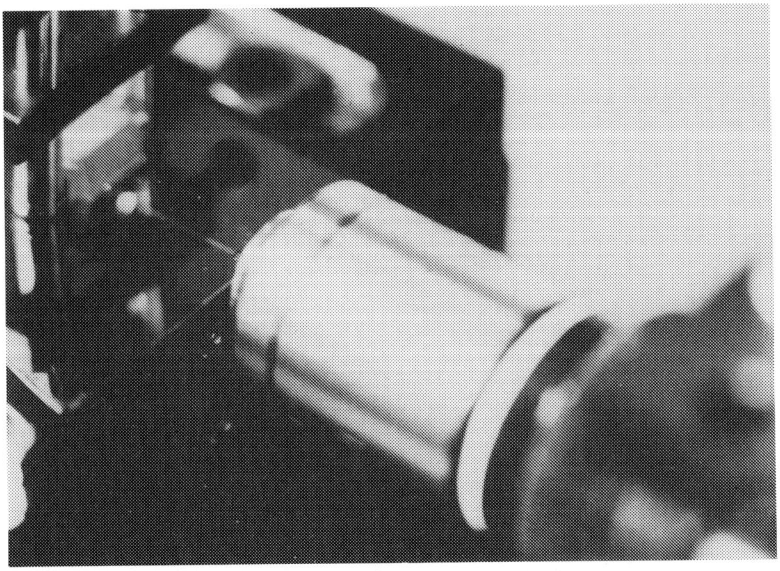

Figure 8.10 *The near-field measurement setup. The He-Ne laser is used for alignment for long-wavelength measurements. The photograph shows a close-up of the substrate with the prism and the microscope objective.*

8.4.4 Lateral field profiles

Lateral field profiles are measured by coupling light into a film or a waveguide with a prism and imaging the near field at a cleaved end face onto a CCD-camera (or an IR-vidicon for the long wavelengths) with a microscope objective. The image is processed by a video-digitizer card installed in a personal computer. For this type of measurement a second measurement setup was developed, analogous to the first, but with a rotating laser mount and a fixed substrate in order to avoid the necessity of mounting the camera system on a rotating substrate holder. The setup is shown in figure 8.10.

With this measurement setup the intensity pattern through a component can be followed by appropriately positioning a number of identical devices onto the substrate as illustrated in figure 8.11. The mask pattern is aligned relative to a natural cleave, then processed and cleaved again after being scratched at the position indicated with a marker. The second cleave is required because the circuit quality at the substrate edge is poor due to build up of the photoresist during spinning. If the tapers are located as indicated in the figure, the intensity profile through the taper region can be measured by exciting them one by one. At the top of the figure three oscilloscope images show the video lines registered across the lateral intensity distribution at the beginning, the center, and the end of a taper.

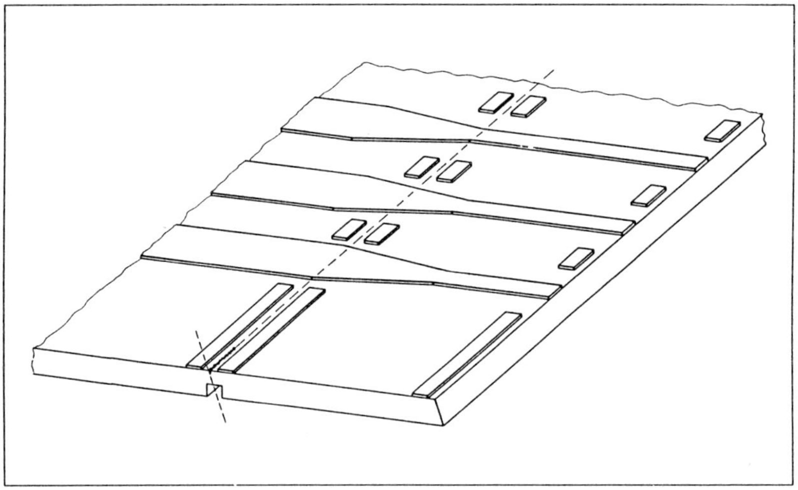

Figure 8.11a *Circuit design (schematic) for measuring the field through a component. The cleave initiated at the marker (dashed line) will pass through the beginning, the center, and the end of the three tapers, respectively.*

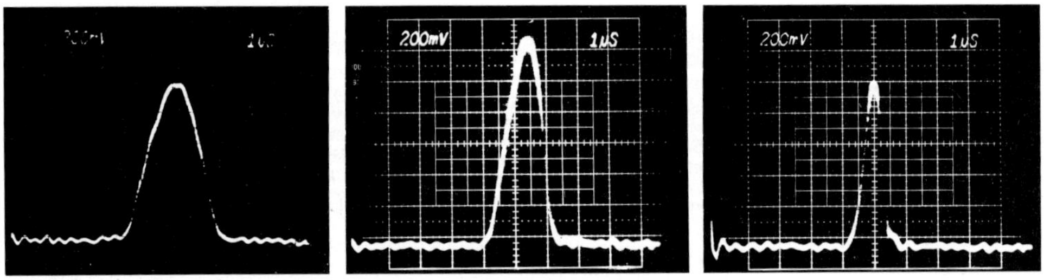

Figure 8.11b *Three oscilloscope images showing the video lines recorded across the lateral intensity distributions measured at the three positions shown in figure 8.11a.*

Figure 8.12 demonstrates the potential quality of the intensity measurement. The modal intensity distribution of the TE_0-mode in a 5 µm wide waveguide, both computed (with the effective index method) and measured using the above setup is shown. It was found that the agreement between theory

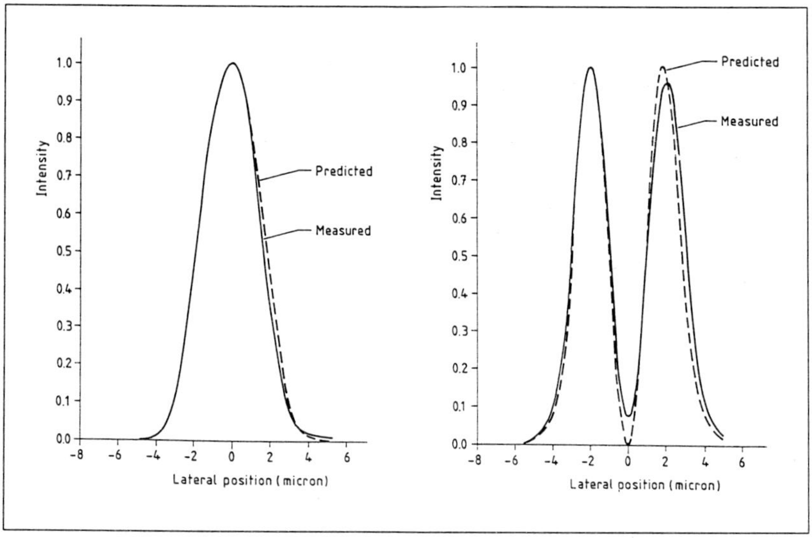

***Figure 8.12** Calculated and measured lateral intensity profile of a* TE_0*- and a* TE_1*-mode in a 5 μm waveguide.*

and experiment is optimal for monomode waveguides; small imperfections in multi-mode guides or waveguide slabs lead to the excitation of unwanted higher-order modes or field perturbations. Furthermore, the quality of the cleave is important, especially if a microscope objective with a small numerical aperture which does not capture all of the radiation coming out of the waveguide structure is employed.

Cleaving of silicon wafers is more complicated than cleaving GaAs or InP because of the greater strength of the silicon crystal. In appendix 8B the procedure and equipment which we developed for inducing good-quality cleaves at well controlled positions is discussed.

8.4.5 Insertion loss and optical transmission

Two methods have been developed for measuring the optical transmission of planar components, which are more or less complementary. The first method is based on comparison of the guided power before and behind the component by sliding the prism over the component. The second method is based on comparison of the transmission through the component under test with the transmission through a straight reference channel. This can be done with a fixed output prism or by measuring the intensity at a cleaved end face.

A. Measurement with a sliding output prism

If the output prism is slid over a lossy component a step should be detected at the position of the component. If the component has waveguides as input and output ports, and the detected power is recorded logarithmically against the translation, the insertion loss follows directly as the offset between the straight lines recorded before and behind the component. Figure 8.13 gives an example of recordings measured on a series of tapers with different taper angles. As can be seen from the figure, this method is suitable for detecting insertion losses in the order of 0.1 dB. The measured results agree well with theoretical results computed using the Beam Propagation Method (see also van Vaalen and Smit [11]).

The substrate has to be covered with a metallic coating in which slits are etched above the waveguides, as depicted in figure 8.14, for this method to

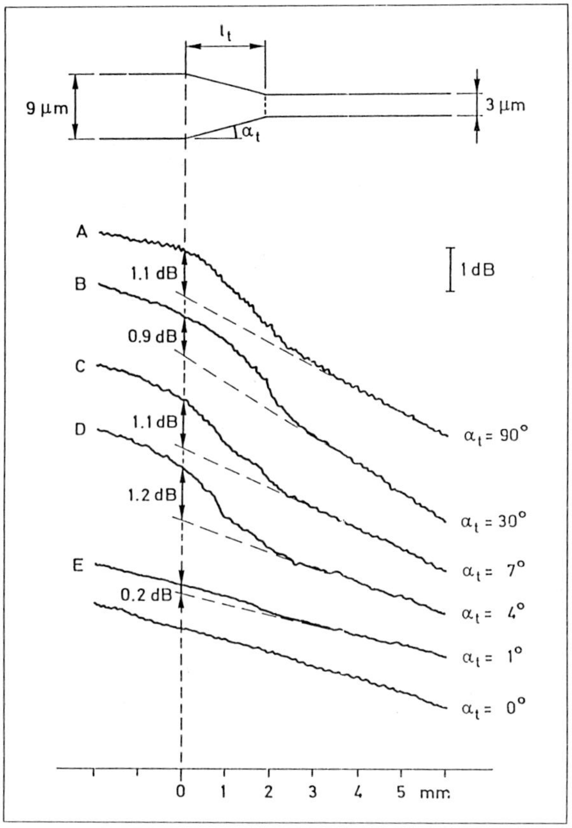

Figure 8.13 *Intensity patterns recorded with a sliding output prism along waveguides containing a taper section. The taper angles are indicated at the right of the curves.*

be employed succesfully. If it is not applied the detected step will be too small or, if the loss of the component is caused by radiation into the slab guide, no step will be detected at all. If the discontinuities in the waveguide pattern are not very abrupt the divergence of the radiation loss will be small, and all of it will arrive at the output prism. The coating shields the output prism from radiation propagating in the film next to the waveguide.

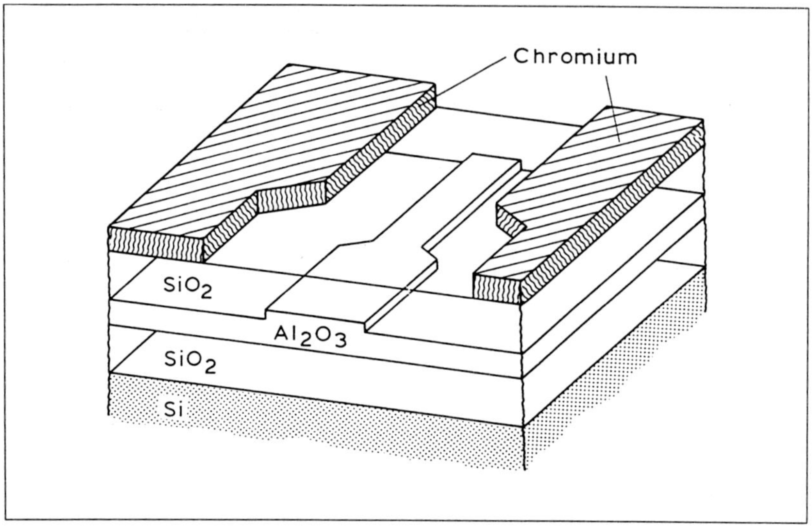

Figure 8.14 Planar taper with chromium coating to prevent lateral radiation from being detected by the output prism.

B. Measurement at a cleaved end face

The near-field measurement described in the previous paragraph is also suitable for comparative measurement of a number of components positioned parallel to each other, or of components with several output channels, such as directional couplers, wavelength demultiplexers and polarization splitters.

An attractive measuring method involves excitation of a number of closely spaced waveguide channels simultaneously with a broad input beam. Examples of results measured in this way can be found in chapters 5 and 6 (figures 5.8 and 6.33). Under appropriate conditions (clean surfaces, optimal beam alignment, and end face of good quality) the uniformity of the measured throughput is within ± 0.5 dB. The accuracy of this method is, thus, less than the sliding-prism method. A great advantage, however, is that all relevant channels can be observed simultaneously, which greatly enhances the ease of measuring.

C. Measurement with a fixed output prism

Similar measurements can be performed with a fixed output prism. In this approach the channels have to be excited one by one because the individual channels cannot be separated with the detector diode glued onto the prism. This can be done by using a (weakly) focused input beam ($w_o \simeq 20\ \mu m$), and (transversely) translating the whole platform (including the prisms) relative to the input beam, thus holding the input coupling conditions as reproducible as possible. The accuracy of this type of measurement is slightly better than that of measurement at a cleaved end face because of the absence of the cleaved end face which can affect the measured throughput. The coupling reproducibility at the output prism is good, because the coupling efficiency is close to 100%. Pennings [12] employed this method for measuring waveguide bend losses.

8.5 Conclusions

Prism coupling is a versatile and non-destructive technique for measuring transmission properties of planar optical films, components, and circuits. The application of a weakly focused "oversized" input beam has been shown to increase the selectivity of exciting single modes, and to reduce the sensitivity to environmental vibrations. Measurement accuracies in well-designed experiments are in the order of 0.1 to 0.5 dB, depending on the specific method employed.

Appendix 8A Prism coupling efficiency for a three-dimensional input beam.

In this appendix expressions will be derived for calculating the three-dimensional overlap-integral (equation 8.13), which has been used for computing the results shown in figure 8.5:

$$\eta = \frac{|\iint U_i^*(y,z)\, U_e(y,z)\, dydz|^2}{\iint |U_i(y,z)|^2 dydz \iint |U_e(y,z)|^2 dydz}. \qquad (13)$$

In this formula $U_i(y,z)$ is the *y-z* dependence of the input beam at the bottom of the prism, and $U_e(y,z)$ describes the *y-z* dependence of the exponentially increasing mode which is required for 100% coupling efficiency. The input beam is assumed to be Gaussian and described by:

$$U_i(\xi,\chi,\zeta) = G_t(\xi,\zeta)\, G_l(\chi,\zeta)\, exp(-jn_p k_o \zeta), \qquad (18)$$

where ξ, χ and ζ are the transverse, the lateral and the longitudinal coordinates in a cartesian coordinate system where the ζ-axis coincides with the beam axis, as depicted in figure 8.15, and:

$$G_t(\xi,\zeta) = exp(-\xi^2/w_\xi^2), \qquad (18a)$$

$$G_l(\chi,\zeta) = \{\frac{w_{\chi o}}{w_\chi(\zeta)}\}^{1/2} \; exp\left[-\frac{\chi^2}{w_\chi^2(\zeta)} - \tfrac{1}{2}j\frac{n_p k_o \chi^2}{R_\chi(\zeta)} + \tfrac{1}{2}j\, arctan(\frac{\zeta}{b_{\chi o}})\right], \qquad (18b)$$

$$w_\chi(\zeta) = w_{\chi o}\; \{1 + (\frac{\zeta}{b_{\chi o}})^2\}^{1/2}, \qquad (18c)$$

$$R_\chi(\zeta) = \zeta\,\{1 + (\frac{b_{\chi o}}{\zeta})^2\}, \qquad (18d)$$

$$b_{\chi o} = \tfrac{1}{2} n_p k_o w_{\chi o}^2 . \tag{18e}$$

In these formulae G_t and G_l describe the transverse and the lateral amplitude profile, w_ξ and w_χ are the transverse and the lateral $1/e^2$ half widths, R_χ describes the lateral phase-front curvature, and $b_{\chi o}$ is the lateral focal depth. Our notation follows that of Kogelnik [13].

The transverse beam waist w_ξ will normally be such that the ζ-dependence of the transverse beam shape (equation 8.18a) is negligible in the coupling region. In the formula for the lateral beam shape description (equation 8.18b) the term between brackets accounts for the amplitude reduction due to the broadening of the beam. In the exponent the first term describes the Gaussian beam profile and the second term the spherical phase fronts. The third term disappears at $\zeta=0$ and introduces a $\pm\frac{1}{4}\pi$ phase shift for large values of ζ.

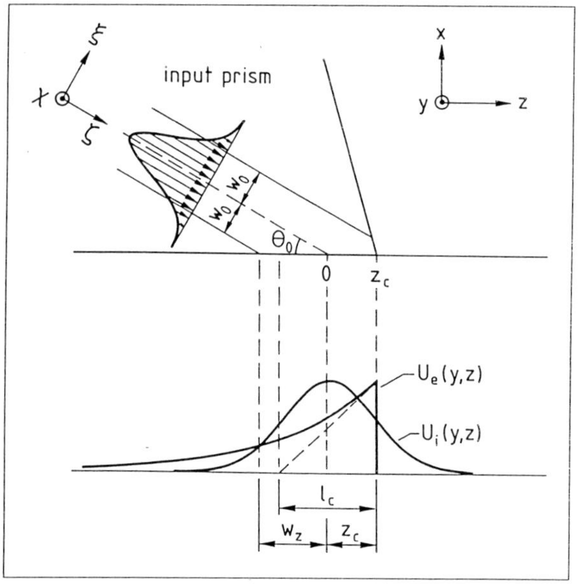

Figure 8.15 *Three-dimensional coupling geometry.*

For the transverse coupling problem, optimum coupling occurs if the beam focus coincides with the intersection point with the bottom of the prism. We will assume that the same holds for the lateral focus. Although beam shaping in two dimensions might shift the optimal focal positions, it is believed that the effects on the following analysis will be small.

From equation 8.18 the following expression for $U_i(y,z)$ in the y-z plane (i.e. at the bottom of the prism) is derived:

$$U_i(y,z) = G_z(z)\, G_y(y,z)\, exp(-jn_p k_o z cos\theta), \tag{19}$$

where

$$G_z(z) = exp(-z^2/w_z^2), \tag{19a}$$

$$G_y(y,z) = \{\frac{w_{yo}}{w_y(z)}\}^{1/2}\; exp\left[\frac{-y^2}{w_y^2(z)} - \frac{1/2 n_p k_o y^2}{R_y(z) cos\theta} + 1/2 j\, arctan(\frac{z}{b_{yo}})\right], \tag{19b}$$

$$w_z = w_\xi / sin\theta, \tag{19c}$$

$$w_y(z) = w_{yo}\; \{1+(\frac{z}{b_{yo}})^2\}^{1/2}, \tag{19d}$$

$$R_y(z) = z\,\{1+(\frac{b_{yo}}{z})^2, \tag{19e}$$

$$w_{yo} = w_{\chi o}, \tag{19f}$$

$$b_{yo} = 1/2\, \frac{n_p k_o}{cos\theta} w_{yo}^2. \tag{19g}$$

In this formula G_y and G_z describe the y- and z-dependence of the amplitude profile in the y-z plane, w_y and w_z are the lateral and longitudinal $1/e^2$ half widths in the same plane, $R_y(z)$

describes the lateral phase-front curvature, w_{yo} is the lateral beam waist, and b_{yo} the lateral focal depth.

For the field $U_e(y,z)$, which is required for 100% coupling efficiency, the y-dependence is described by the fundamental mode profile $U_m(y)$. To simplify the expressions we will approximate the mode profile as a (non-diverging) Gaussian beam $exp(-y^2/w_m^2)$. Using these approximations we find:

$$U_e(y,z) = exp\,(-y^2/w_m^2 + (z-z_c)/l_c - j\beta z), \qquad z \le z_c,$$
$$= 0, \qquad z > z_c, \qquad (20)$$

in which l_c is the prism coupling length governing the z-dependence, and z_c the position of the end of the input prism. Substitution of eqs. 8.19 and 8.20 (with $\beta = n_p k_o cos\theta_o$) into the integrals of equation 8.13 yields:

$$\int\int |U_i(y,z)|^2\, dydz = \tfrac{1}{2}\pi\ w_{yo} w_z, \qquad (21)$$

$$\int\int |U_e(y,z)|^2\, dydz = \tfrac{1}{2}\, w_m l_c \sqrt{\pi/2}\,, \qquad (22)$$

$$\int\int U_i(y,z)\ U_e^*(y,z)\, dydz = \sqrt{\pi} \int_{-\infty}^{z_c} Q(z)\{\frac{w_{yo}}{w_y(z)}\}^{1/2} exp\,\{-z^2/w_z^2 + (z-z_c)/l_c + \tfrac{1}{2} j\, arctan(\frac{z}{b_{yo}})\}, \qquad (23)$$

in which

$$Q(z) = w_m w_y(z)/\{w_m^2 + w_y^2(z) + j\, w_m^2 z/b_{yo}\}^{1/2}, \qquad (24)$$

Substitution of eqs. 8.21, 8.22 and 8.23 into equation 8.13 leads to an expression for η containing a single integration in z, which is easily computed numerically.

Appendix 8B Substrate cleaving.

Cleaving silicon substrates in a well controlled manner is more complicated than cleaving GaAs and InP substrates because of the greater strength of silicon crystals. Silicon crystals cleave along {111}-planes. In {110}-oriented wafers the {111}-planes are positioned perpendicular to the wafer surface, so that cleaved {110} wafers show orthogonal cleaving facets. The more customary {111} orientation yields cleaves at 70.4° with the wafer surface. {100}-oriented wafers are not suitable for cleaving purposes. Figure 8.16 shows the orientation of the cleaves for different wafer types.

We developed a cleaving apparatus based on the same principles as existing GaAs cleaving equipment. Figure 8.17 illustrates the operation. The substrate is mounted between two pieces of Mylar foil. After making a small scratch through a window in the foil, in order to fix the starting point of the cleave, a small circular rod, which is positioned

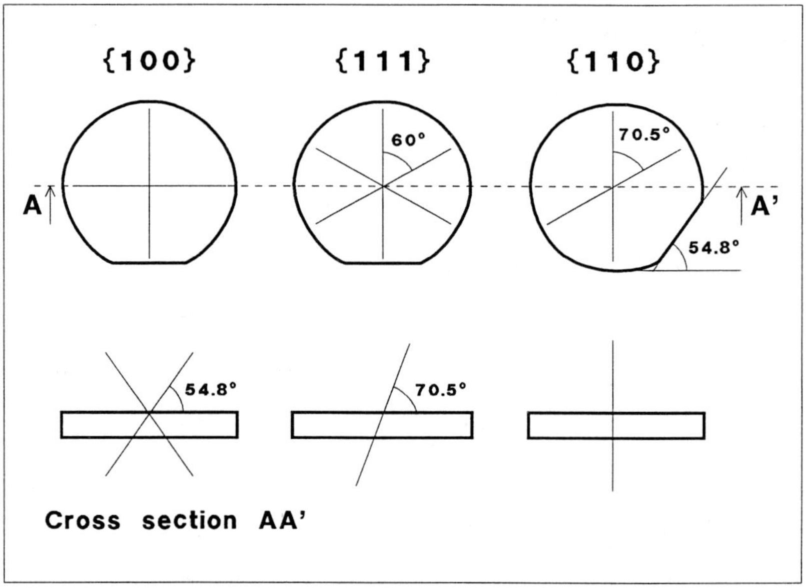

Figure 8.16 *Orientation of the cleaves for different wafer types. Thin solid lines indicate the orientation of the {111} cleaving planes.*

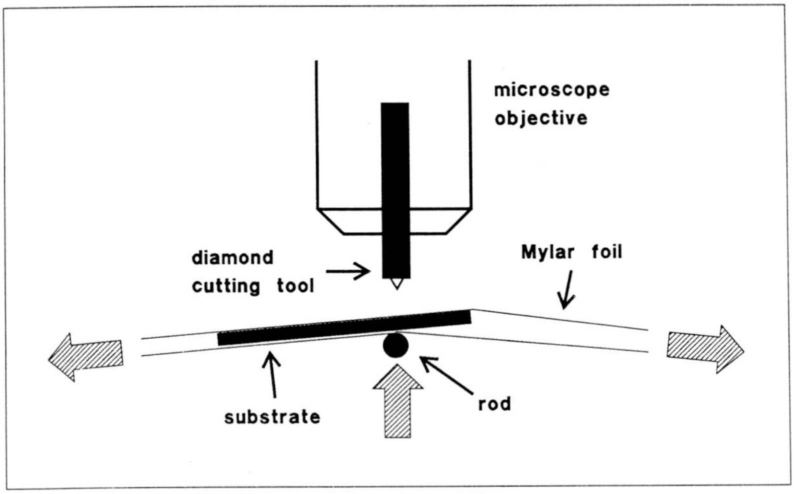

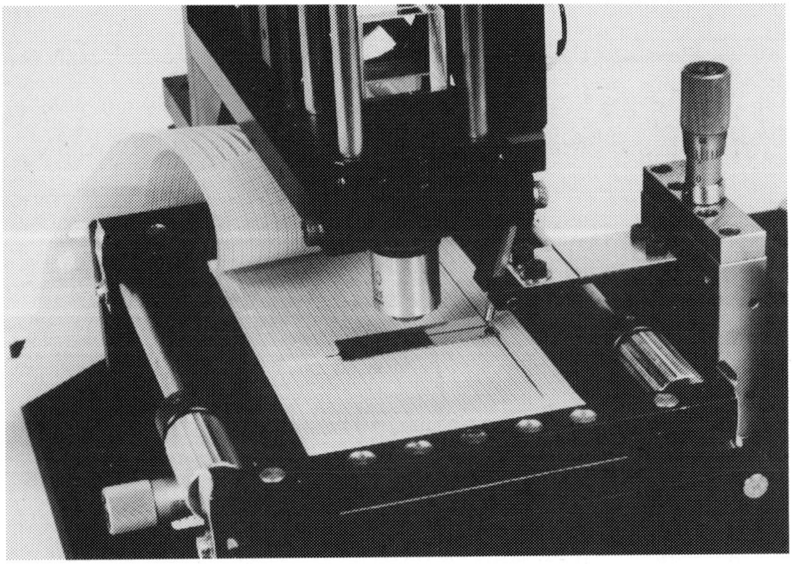

Figure 8.17 *Upper figure: Principle of the scratching and cleaving method. Lower figure: Close-up of the scratching and cleaving section of the developed instrument.*

under the scratch and parallel to the natural cleaving direction, is pressed against the back side of the substrate with increasing force, until it is cleaved. The positioning of the scratch is controlled with a microscope.

Wafers with thermal oxide films of 4 μm are very difficult to cleave in a controlled way because of the compressive stress in these films which impedes cleave propagation. For these wafers dicing and polishing yields better results.

References

8.1 P.K. Tien, R. Ulrich and R.J. Martin, "Modes of propagating light waves in thin deposited semiconductor films", Appl. Phys. Lett., Vol. 14, pp. 291-294, 1969.

8.2 P.K. Tien, R. Ulrich, "Theory of prism-film coupler and thin-film light guides", J. Opt. Soc. Am., Vol. 60, pp. 1325-1337, 1970.

8.3 R. Ulrich, "Theory of prism-film coupler and thin-film light guides", J. Opt. Soc. Am., Vol. 60, pp. 1337-1350, 1970.

8.4 H.G. Unger, "Planar optical waveguides and fibres", Oxford: Clarendon Press, section 2.7, 1977.

8.5 C.A. Millar, R.H. Hutchins and P.J.R. Laybourn, "Modes of a 3 dimensional sandwich-ribbon optical waveguide", Microw. Opt. and Acoust., Vol. 1, pp. 27-33, 1976.

8.6 P. Simova, I. Savatinova and L. Tsonev, "Propagation characteristics of Ag^{+} ion-exchanged stripe waveguides", Appl. Phys., Vol. 22, pp. 237-240, 1980.

8.7 D.K. Wang and P.J.R. Laybourn, "Coupling between prism and rectangular optical waveguide", IEE Proc., Vol. 130, Pt. H., pp. 255-260, 1983.

8.8 W.A. Pasmooij, M.K. Smit and P.A. Mandersloot, "Prism-coupling of light into narrow planar optical waveguides", Journal of Lightwave Technology, Vol. 7, pp. 175-180, 1989.

8.9 M.K. Smit, C.J. van der Laan and G.A. Acket, "Aluminium oxide films for integrated optics", Thin Solid Films, Vol. 138, pp. 171-181, 1986.

8.10 M.K. Smit, "Sputtered planar optical waveguides", Sensors & Actuators, Proc. S & A symp. of the Twente Univ. of Techn., Enschede, the Netherlands, Oct. 30-31, 1986.

8.11 M.J.M. van Vaalen and M.K. Smit, "Fabrication and measurement of low-loss planar optical tapers", Fibre and Integrated Optics, Vol. 7, pp. 235-239, 1988.

8.12 E.C.M. Pennings, "Bends in Optical ridge waveguides", Ph.D Thesis, Delft University of Technology, Delft, 1990.

8.13 H. Kogelnik, "Imaging of optical modes", Bell. Syst. Tech. J., Vol. 44, pp. 455-494, 1965.

Chapter 9

A novel approach to fiber coupling

The problem of low-cost coupling of planar optical circuits with small waveguide dimensions to fibres has not yet been solved. In most approaches micro-optical components are used to adapt the mode profiles of both waveguides to each other. In this chapter a novel method is decribed for realizing this adaption without the use of micro-optics.

9.1 Introduction

One of the most important problems with respect to the application of integrated optical circuits is their connection to optical fibers. Low-cost solutions, based on direct coupling, suffer from high coupling losses [1]. To obtain low losses the planar mode profile must be matched to the fibre mode. This is most frequently done employing micro-optical components [2-5]. Alignment tolerances are in the submicron range due to the dimensions of the planar waveguide. The coupling problem is further aggravated by the strong ellipticity of the mode profile within most small-size waveguides.

Although efficient coupling methods are presently employed industrially, they suffer from high fabrication costs. The cost of a "pigtailed" cleaved-mirror laser, for example, is many times the cost of the laserchip. It will be difficult, if not impossible, to reduce the coupling costs to the same order as envisioned chip costs. The main problems are in the alignment and the packaging, where relative positions have to be fixed with sub-micron precision under widely varying operating conditions.

The problem can be solved by creating a gradual local change of the planar waveguide geometry near the coupling region in order to adapt the mode profile to the fibre mode [6,7]. In this approach the alignment and packaging costs would be reduced. The remaining problem is the one of reproducible fabrication of the non-planar films which are necessary for adiabatic conversion of the mode profile. In the present paper an approach is proposed and experimentally tested which allows for a fibre-matched output beam which does not require adaption of the transverse waveguide geometry. Results were recently published [8].

9.2 Basic principle

The prism coupler is widely employed for experimental purposes. Coupling efficiency of the commonly employed configuration which has a uniform tunneling gap is maximally 80% for Gaussian beams [9]. Better coupling efficiencies can be obtained by employing a tapered tunneling layer [10] which yields a bell-shaped output beam, as shown in figure 9.1.

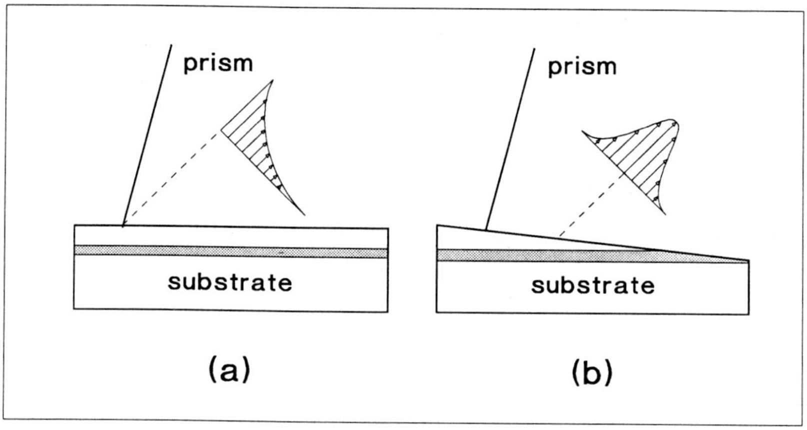

Figuur 9.1 Geometry of the beam radiating from a waveguide covered with a uniform tunneling layer (a) and a tapered layer (b).

The width of the output beam is determined by the taper angle of the gap, and will decrease monotonously to the transverse width of the guided mode if the taper angle is increased to 90°, corresponding to a perpendicular endface. A particular taper angle will thus exist for which the transverse width of the outgoing beam is matched to the width of a fibre mode. Consequently, it is possible to match the transverse width of the output beam of an arbitrary planar waveguide to a fibre mode by choosing this angle appropriately. The lateral width is determined by the lateral mode width; it is easily matched to the fibre mode by employing a planar taper.

The principle can be applied to couple light from a planar circuit to a fibre if a high-index liquid is employed between the fibre and the waveguide. The index of this material has to be greater than the effective index of the mode which is to be coupled out of the waveguide. Permanent connections can be made, in principle, using UV-hardening epoxies or low melting-point eutectics, if the optical and mechanical properties can be sufficiently stabilized. In the following we will refer to the high-index material as the "prism material".

The potential of the above coupling principle will be theoretically analysed and experimentally demonstrated below for a wavelength of 633 nm. The latter wavelength was chosen because of the availability of accurate measurement equipment operating at this wavelength. The principle applies equally well to other wavelengths. The present experiments were designed with a view to establishing the validity of the theoretical description given in the next section.

9.3 Coupler analysis

9.3.1 Output beam shape

Two regimes should be distinguished in calculating the shape of the output beam. For small taper angles evanescent field coupling will be the dominant mechanism. For large taper angles the coupling region will become so short that evanescent coupling and diffraction effects are small and the coupling can be described as refraction of a parallel beam through a skew endface.

A. Evanescent coupling

In the evanescent coupling regime the radiated power is equal to the power leakage of the guided mode so that the output beam shape can be computed by solving the differential equation describing the power decay of the guided mode due to radiation leakage:

$$\frac{dP(z)}{dz} = -2\,\alpha(z)\,P(z). \tag{1}$$

The attenuation coefficient α is dependent on the thickness g of the tunneling layer and follows from the effective index as $\alpha(g) = N''(g)\,k_o$, in which $-N''(g)$ is the imaginary part of the effective index N. The complex effective index of a prism-loaded waveguide can be obtained numerically in several ways; we employed a transfer-matrix method [11]. The local attenuation coefficient $\alpha(z)$ follows from $\alpha(g)$ by substituting the (linear) dependence of g on z. Numerical integration of eq. 9.1 yields the z-dependence $P(z)$ of the guided power, from which the power $S(z)$ radiated per µm follows as:

$$S(z) = -\frac{dP}{dz} \simeq 2\,\alpha(z)\,P(z). \tag{2}$$

Figure 9.2 shows the beam shapes $S(z)$ along the z-axis, radiating from the fundamental modes (*TE* and *TM*) in an aluminum oxide waveguide on silicon substrate, as described in section 3, for two different taper angles (1° and 2°).

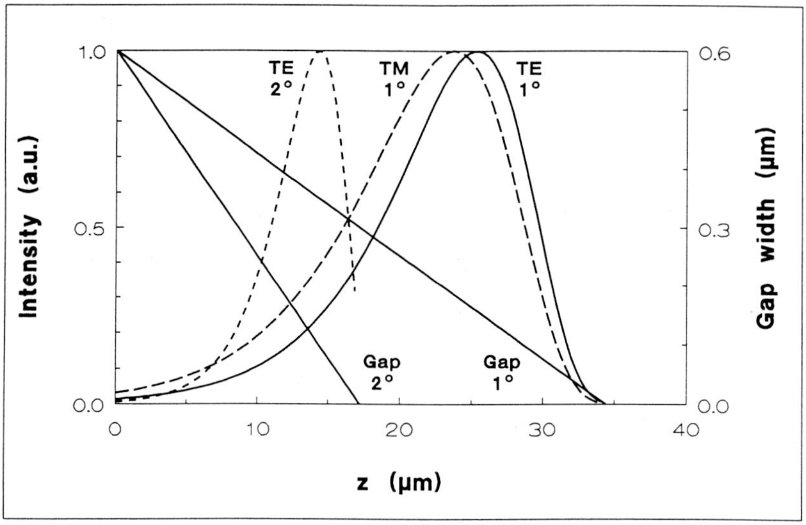

Figure 9.2 *Shape of the beam coupled into the prism medium from a guided mode through a linearly tapered tunneling layer. Beam shapes are shown for 1° and 2° taper angle, for the TE and TM-polarized fundamental modes.*

The above computations are valid so long as the mode is able to adapt its profile adiabatically to the changing waveguide structure. This will be the case so long as the ray propagation angle θ'_m after reflection is smaller than the critical angle at the reflection interface, i.e. $\theta'_m = \theta_m + 2\,\theta_t < \theta_c$, in which θ_c is the critical angle of the waveguide structure and θ_m is the ray propagation angle corresponding to the guided mode. The reason for employing small-sized waveguides is usually a requirement to keep the waveguide single moded despite a relatively large index contrast. Critical angles in small-sized waveguides will therefore be large, in the order of 10° to 20°, and no problems should be expected in applying the above computation method until close to the end of the waveguide where the mode approaches cutoff and θ_m comes close to θ_c.

B. Refraction

In the refraction regime most of the guided power reaches the skew endface before being coupled out of the waveguide by evanescent coupling or diffraction. Computation of the beam shape in the refraction regime is straightforward. From figure 9.2 it can be seen that the transition from the evanescent to the refraction regime begins before taper angles of 2°; not all of the guided power is coupled out of the waveguide before it becomes degenerate (at $z=z_e$ in figure 9.6a) at this taper angle.

9.3.2 Output beam width

In the evanescent coupling regime the effective width w_s of the output beam measured along the z-axis (see figure 9.3) is computed from:

$$w_s = \frac{\int S(z)\,dz}{S_{max}}. \tag{3}$$

Figure 9.4 shows the effective width as a function of the taper angle for both polarizations. In the refraction regime w_s follows directly from the effective mode width w_m (figure 9.3):

$$w_s = w_i \cos\theta_t = w_m / tg\,\theta_t, \tag{4}$$

in which w_i is the beam width measured along the skew interface. The dotted curve in figure 9.4 shows the width computed according to this formula. The

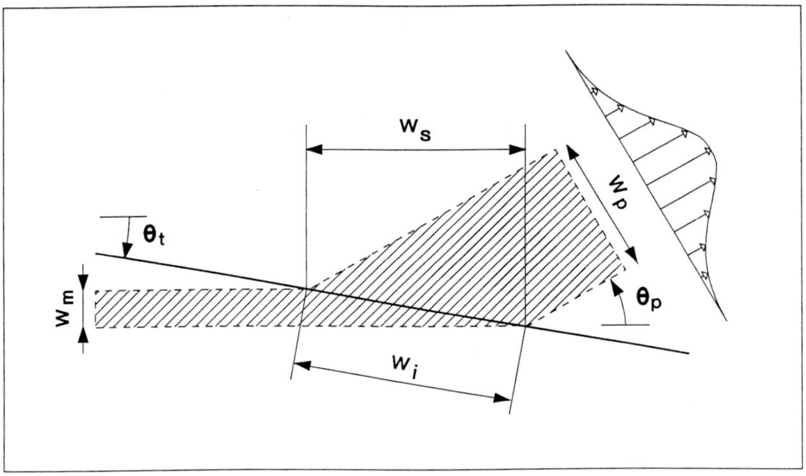

Figure 9.3 *Refraction of a parallel beam through a skew endface.*

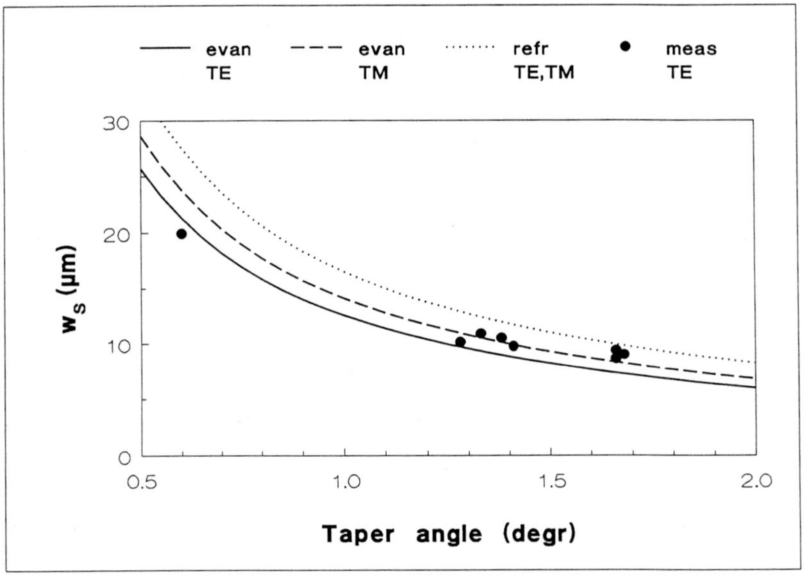

Figure 9.4 *The effective width of the output beam measured along the z-axis as computed from eq. 9.3 (evanescent coupling) for TE- and TM-polarization, and from eq. 9.4 (refraction, both polarizations). The marked points represent experimentally measured data (TE).*

various predictions are not far apart. If we were able to correct for the truncation error which can be observed in the beam shape for the 2° taper (see figure 9.2), the widths found for the evanescent field coupling method would become slightly greater and the curves would come even closer for taper angles in excess of 1.5°.

The effective beam width w_p (see figure 9.3) is derived from the assumption that the amplitude profiles match at the skew interface:

$$w_p = w_i \sin(\theta_p + \theta_t) \simeq w_s \sin\theta_p . \qquad (5)$$

In this formula θ_p is the propagation angle of the output beam which follows, both for the refraction and the evanescent coupling regime from the field continuity conditions at the skew interface:

$$\theta_p = \arccos\left(\frac{N}{n_p}\cos\theta_t\right) - \theta_t \simeq \arccos\left(\frac{N}{n_p}\right) - \theta_t , \qquad (6)$$

N being the mode index. For the waveguide configuration used in the present experiments $w_m \simeq 0.30\ \mu m$, with a negligible difference between the two polarizations. The radiation angle θ_p is approximately 25°. In the refraction regime an effective width of 5 µm, which is representative for a fibre mode, requires a value $w_s \simeq 12.5\ \mu m$, from which a taper angle of 1° is computed using equation 9.4.

9.3.3 Coupling efficiency

Coupling performance is determined by the overall coupling loss L, which consists of two components:

$$L = L_r + L_m, \qquad (dB). \tag{7}$$

L_r represents the reflection losses at the skew interface and at the fibre tip, and L_m the loss due to field mismatch between the output beam and the fibre mode. The latter mismatch may occur in the amplitude as well as in the phase of both fields. The various loss mechanisms will be discussed subsequently.

A. Reflection

Reflection loss at the fibre tip is easily estimated using the Fresnel reflection formula; it can be eliminated by applying an appropriate anti-reflection coating. The reflection at the skew interface is more difficult to predict. In the evanescent coupling regime the mode will be able to adapt itself adiabatically to the changing waveguide geometry and there will be no reflected field. In the refraction regime the loss will be dominated by Fresnel-like reflection which approaches 100% near grazing incidence.

For guided beams, however, the reflection deviates from plane-wave reflection, a difference which increases with increasing incidence angle [12,13]. This mechanism has been analysed for reflection at a plane normal to the propagation direction of the mode. The reflection of a guided wave at strongly skew interfaces has not yet been analysed theoretically. We determined it experimentally, as will be discussed in section 9.5.

B. Amplitude mismatch

The loss L_m due to field mismatch can be computed by evaluating the overlap integral between the output beam and the fibre mode along the *z*-axis. For ease of computation we approximated the fibre mode as a Gaussian beam. For the experimental waveguide configuration and a taper angle of 1° the effective width of the *TE*-polarized beam was found to be 12.5 μm and that of the *TM*-polarized beam 13.5 μm, as can be seen from figure 9.4. Figure 9.5 shows the overlap of the (square root of the) corresponding beam profiles *S(z)* with a Gaussian profile with 12.5 μm width, as a function of the *z*-coordinate of the Gaussian beam centre.

From the figure it can be seen that on proper alignment of the beam relative to the output beams a transverse coupling efficiency in excess of 94% can be obtained for the *TE*- and *TM*-polarized beams simultaneously. This corresponds to a 0.25 dB coupling loss to a Gaussian beam; coupling loss to a fibre mode is expected to be close to this value.

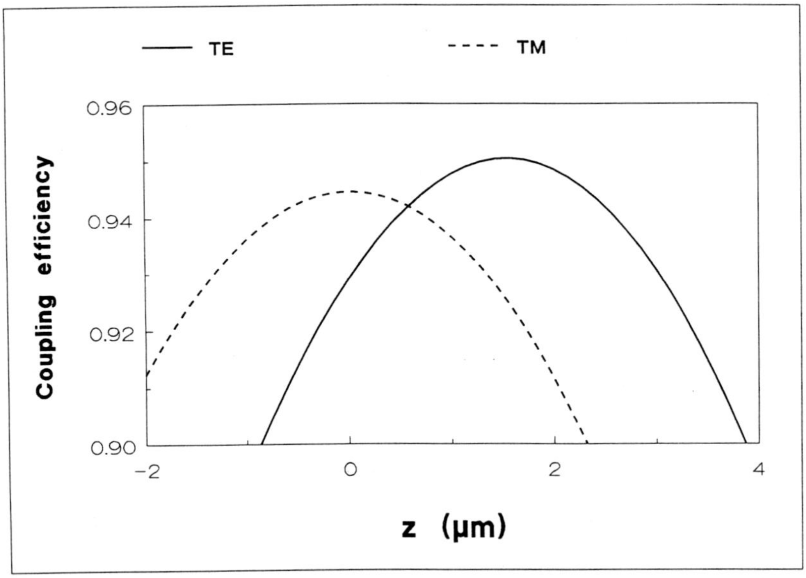

Figure 9.5 Coupling efficiency from the output beam for a 1° taper angle to a Gaussian beam with 12.5 µm effective width, measured along the z-axis.

-Curve 1: TE-polarized beam, 12.5 µm effective width
-Curve 2: TM-polarized beam, 13.5 µm effective width

C. Phase mismatch

Because the two polarizations have different propagation constants it is not possible to match their phases simultaneously. The resulting coupling loss can be estimated by comparing the difference in radiation angles within the prism material with the numerical aperture of the fibre.

In the aluminum oxide waveguide system the angular difference between the two polarizations in a medium with $n=1.74$ is in the order of 1½°, both at short and long wavelengths (assuming the film thickness is adapted so as to hold the *V*-parameter at a fixed value of 2, see chapter 3). If the fibre mode is approximated as a Gaussian beam, the effective numerical aperture[1] of the fibre follows from $NA_e = \lambda/(4w_e)$. For an effective fibre mode width $w_e \simeq 5\ \mu m$ at 1.3 μm wavelength the numerical aperture will be 2°. If the fibre alignment is centered between the radiation angles of the two polarizations, the angular misalignment will amount to 40% of the numerical aperture for each polarization. Assuming a Gaussian radiation pattern, an angular misalignment equalling the effective Numerical Aperture angle will cause a coupling loss of approximately 3 dB (for Gaussian beams the effective width comes close to the FWHM). Because the coupling losses reduce quadratically with decreasing misalignment angle, they are estimated to be in the order of 0.5 dB.

In conclusion, the anticipated excess loss due to both amplitude and phase mismatch is expected to be well below 1 dB for the present experimental waveguide system. As the output beam closely resembles the fibre mode, the loss due to residual misalignment should be the same as in the fibre-to-fibre coupling case.

---

1) The effective numerical aperture is related to the Gaussian beam divergence parameter θ_o according to (see also chapter 3, section 3.2.1B):

$$NA_e = \tfrac{1}{2}\theta_o\sqrt{(\pi/2)}.$$

Using the relations $\theta_o=\lambda/(\pi w_o)$ and $w_o=w_e\sqrt{(2/\pi)}$ we find the following relation between NA_e and w_e:

$$NA_e = \lambda/(4w_e).$$

9.4 Experimental results

To test the model description, developed in the previous section, we fabricated a series of waveguides[1], as depicted in figure 3.3, and provided them with a tapered end face, as shown in figure 9.1b, in the following way. The end section of the waveguides was covered with an additional 0.4 μm SiO_2. In this thick cover region we polished a spherical hole with a metal ball, as indicated in figure 9.6a. Figure 9.7 shows an interference-contrast micrograph of the tapered edge with a waveguide ending up into the taper region.

The taper angle θ_t at the edge of the hole follows from the diameter of the hole, which can be measured using a microscope to be:

$$\theta_t = arcsin\,(\sqrt{D_h^2 - \Delta_o^2}\;/\;D_b) \qquad (8)$$

in which D_h is the hole diameter, D_b the diameter of the polishing ball, and Δ_o the offset of the waveguide axis relative to the hole center.

To determine the quality of the beam coupled out of the waveguide, we coupled light into the waveguide at the thin cover region and imaged the light distribution coupled out of the tapered region onto a CCD camera through

1) Waveguide geometry as depicted in figure 3.3: film thickness $d = 0.25\ \mu m$, ridge height $\Delta d = 0.04\ \mu m$, cover layer thickness $d_{cover} = 0.6\ \mu m$ and waveguide width w=6 μm (adapted to the fiber mode width).

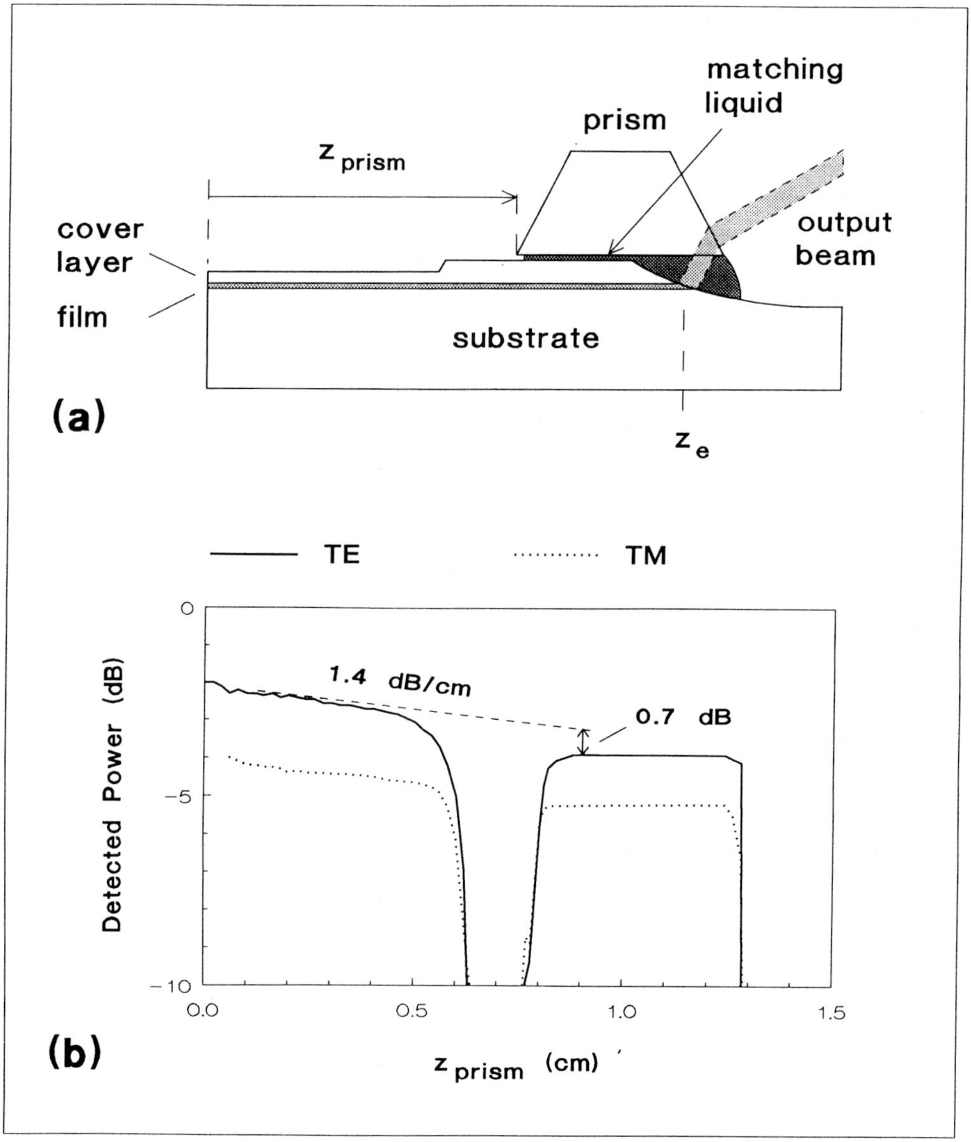

Figure 9.6a *Experimental configuration for verification of the coupling principle. Light is coupled into the waveguide with an input prism at the right side of the substrate (not shown).*

Figure 9.6b *Measurement recording of the power detected by a prism sliding from the taper region towards the input prism. (scale not calibrated.)*

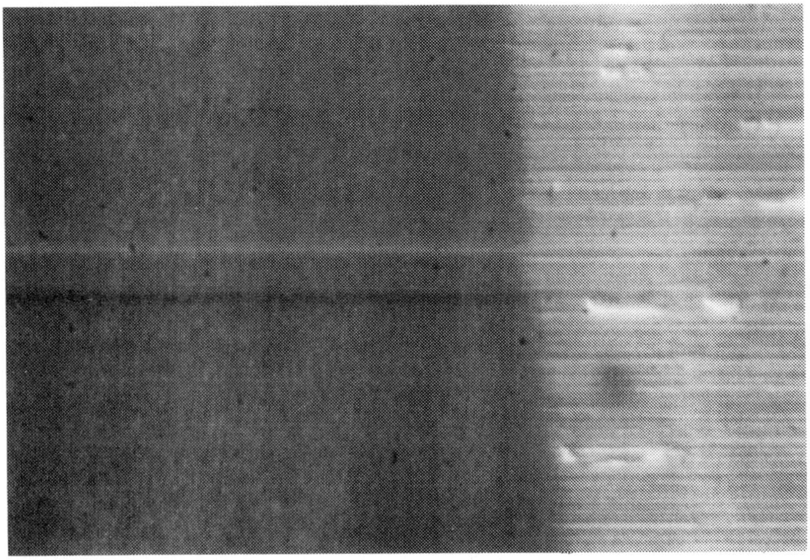

Figure 9.7 Interference-contrast micrograph of a tapered waveguide section, revealing polishing grooves. The waveguide at the left is 6 μm wide.

a small high-index prism with a long-working-distance microscope objective. The prism (SF6 glass, $n \simeq 1.74$) was provided with good optical contact to the waveguide structure using a high-index liquid (CH_2I_2, $n \simeq 1.74$). Figure 9.8a gives an example of the transverse and the lateral light distributions measured with this setup, figure 9.8b of the transverse profiles measured for the *TE*- and the *TM*-polarized mode. The widths of the measured transverse distributions are indicated in figure 9.4, showing good agreement between theory and experiment. Results shown are for *TE*-polarized light.

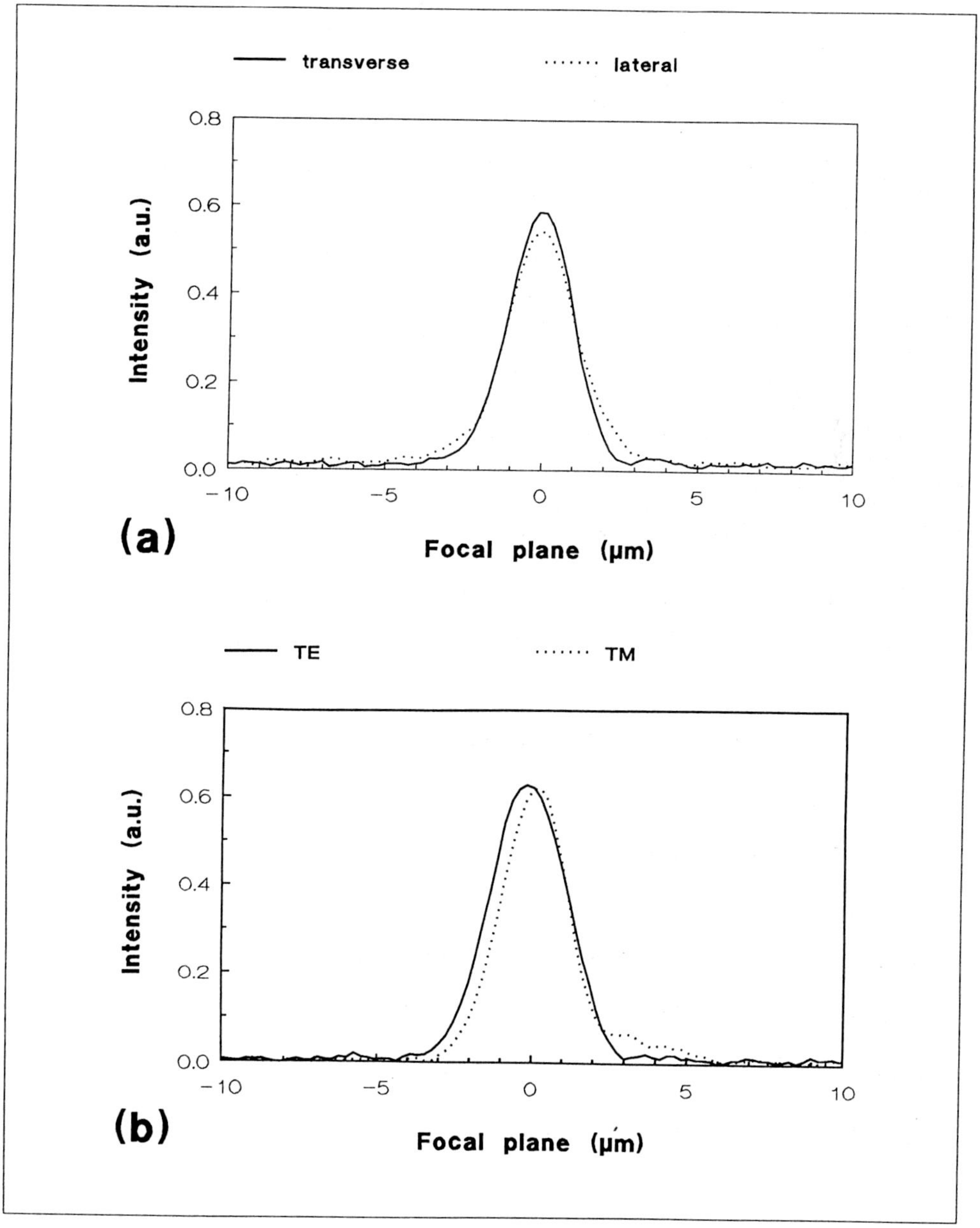

Figure 9.8a *Transverse and lateral intensity distribution at the coupling region, as imaged onto the CCD camera.*

Figure 9.8b *Transverse intensity distribution radiating from the TE and TM-polarized fundamental modes.*

To determine the reflection loss L_r at the skew interface we slide the output prism in the direction of the input prism and record curves as shown in figure 9.6b for taper angles ranging from 1.1 to 1.6°. From left to right the flat zone corresponds to the region where the output beam is detected by the photodiode mounted on the prism. The dip at the left of this zone corresponds to the region where the output beam is no longer detected by the prism-mounted diode. The zone farthest to the left corresponds to the region with a thin cover layer where the guided power is coupled out of the waveguide in the usual way. The slope is due to waveguide attenuation. By comparing the extrapolated curve with the power level in the flat zone at the right, the reflection loss at the skew interface is seen to be less than 1 dB for both polarizations. This is a very promising result. The low reflection loss indicates that evanescent coupling is still the dominant mechanism for taper angles up to 1.6°.

9.5 Conclusions

A novel method is proposed for coupling monomode fibres to planar optical circuits with small waveguide dimensions. It combines the alignment tolerance of butt-coupling with good coupling efficiency, without the need to apply micro-lenses. The method applies to a variety of waveguide structures.

If polishing is performed employing a flat polishing process on a number of circuits parallel to each other, simultaneous coupling to a number of circuits seems feasible, thus opening the way to potentially low-cost production. The experimental results obtained with the surface quality as shown in figure 9.7 indicate that the polishing requirements are not very critical.

Based on theoretical calculations supported by experimental results, coupling losses are anticipated to be in the order of 1 to 2 dB (1 dB beam-shape mis-match loss and reflection losses in the order of ½ dB).

References

9.1 L.A. Reith and J.W. Mann, "Design parameters for a 1.3 μm connectorised low-cost laser package for subscriber loop using graded-index lenses", Electron. Lett., 1987, 23, pp.520,521.

9.2 H. Karstenstein and R. Frankenberger, "High-efficiency two lens laser diode to single-mode fibre coupler with a silicon plano convex lens", J. Lightwave Technol., 1989, 7, pp. 244-249.

9.3 H. Honmou, R. Ishikawa, H. Ueno, and M. Kobayashi, "1.0 dB Low-loss coupling of laser diode to single-mode fibre using a planoconvex graded-index rod lens", Electron. Lett., 1986, **22**, pp. 1122,23.

9.4 H.M. Presby and A. Benner, "Bevelled-microlensed taper connectors for laser and fibre coupling with minimal back-reflections", Electron. Lett., 1988, **24**, pp. 1162,63.

9.5 H.M. Presby, N. Amitay, R. Scotti, and A.F. Benner, "Laser-to-fibre coupling via optical fibre up-tapers", J. Lightwave Technol., 1989, **7**, pp. 274-278.

9.6 R.A. Pattie and M.W. Austin, "Fabrication of tapered couplers in GaAs/GaAlAs rib waveguides", Electon. Lett., 1988, **24**, pp.284,285.

9.7 D.E. Bossi, W.D. Goodhue, and R.H. Rediker, "Reduced-confinement GaAlAs tapered waveguide antenna", IGWO, 1989, pp. 80-83

9.8 M.K. Smit and A.H. de Vreede, "A novel method for matching the mode profile of planar waveguides to single-mode fibers", to appear in Applied Optics, May, 1991.

9.9 H.-G. Unger, Planar optical waveguides and fibres (Oxford: Clarendon Press, 1977).

9.10 R. Ullrich, "Optimum excitation of optical surface waves", J. Opt. Soc. Am., 1971., **61**, pp. 1467-76.

9.11 J. Chillwell and I. Hodgkinson, "Thin-films field-transfer matrix theory of planar multilayer waveguides and reflections from prism-loaded waveguides", J. Opt. Soc. Am. A, 1984, **1**, pp. 742-753.

9.12 J. Buus, "Analytical approximation for the reflectively of DH lasers", IEEE J. Quantum Electron., 1981, **QE-17**, pp.2256,57.

9.13 T. Ikegami, "Reflectivity of mode at facet and oscillation mode in double-heterostructure injection lasers", IEEE J. Quantum Electron., 1972, **QE-8**, pp. 470-476.

Epilogue

Technical science differs from natural science in that its goal is creation rather than analysis and description. It is, therefore, more than an applied science. Its emphasis is not primarily on the extension of analytical methods or descriptions, which is the domain of non-technical science, but on the creative combination of existing analytical knowledge in order to arrive at new technical entities.

By systematically applying the scientific method of analysis and abstraction the desired functional performance of the intended object (in electrical engineering ussually a component, a circuit or a system) is decomposed into a number of more elementary functions, for which standard solutions exist or can be developed. In a second stage the elementary solutions are integrated in a problem-specific way to synthesize the intended technical function. This function is modeled by combining the theoretical descriptions of the elementary functions. For a well-developed technology a single pass through the analysis and synthesis cycle may yield a satisfactory result. In technical research the two stages are often applied iteratively in order to correct for unforeseen differences between modeling and experiments.

In the thesis this approach has been applied to a number of new functional components in the field of integrated optics: a new type of directional coupler, polarization splitters and demultiplexers based on a novel concept, the optical phased array, and a new planar-to-fiber coupling technique. All these components can be described by a combination of three elementary components: waveguides[1], free space[2], and junctions[3]. They were analyzed by combining the well-known properties of the elementary components, and strategies were developed and experimentally tested in order to arrive at optimal designs.

The work reported in this thesis is one step in a continuing process of development which will eventually lead to the achievement of operational designs. The experimental work has been performed using a silicon-based aluminum oxide waveguide system which is well-characterized but has little significance for product development. A next step, which has already been undertaken, is the application of the developed concepts into the more promising InGaAsP/InP material system.

The prospects for such a program are good. The interest in compact and reproducible 3-dB couplers is growing rapidly. The importance of short waveguide bends will increase with increasing circuit complexity. The need for dedicated optical components for use in broadband telecommunication networks has been clearly identified worldwide as a technical challenge, not only for coherent communication systems but also for high-density wavelength division multiplex (HDWDM) systems based on direct detection.

---

1) Straight or bent.
2) Three-dimensional or two-dimensional (planar film).
3) Between different waveguides and between waveguides and free space.

The concepts which have been developed are not restricted to telecommunication applications; they may also be employed succesfully in optical processing and optical sensor technology, as well as in consumer and professional applications. With the work reported in this thesis the laboratory of telecommunication and remote sensing technology has established its potential for participation in international development of photonics.

Literature

E. Schuurman, *Techniek en Toekomst (Technology and the future),* Van Gorcum & Comp. N.V., Assen, The Netherlands, 1972.

Epiloog

De technische wetenschappen verschillen van de natuurwetenschappen doordat ze een scheppend element bevatten. Waar de natuurwetenschappen zich op een bestaande werkelijkheid richten, doen de technische wetenschappen dat op een te vormen werkelijkheid. Ze zijn daarom niet primair gericht op het ontwikkelen van analytische methoden en analytische kennis, die het domein van de niet-technische wetenschappen vormen, maar op het op creatieve wijze combineren van bestaande kennis om daarmee tot nieuwe technische ontwerpen te komen.

Door het systematisch toepassen van de generaliserende wetenschappelijke methode van analyse en abstractie wordt de gewenste functie van het te maken object (in de elektrotechniek meestal een systeem, een schakeling of een component) opgedeeld in een aantal meer elementaire deelfuncties, waarvoor standaardoplossingen bestaan of kunnen worden ontwikkeld. In een volgende fase worden deze elementaire oplossingen op een individualiserende (op de specifieke functie toegespitste) wijze geïntegreerd om tot de gewenste functie te komen. Deze functie kan worden beschreven door de theoretische beschrijvingen van de elementaire deeloplossingen te combineren. Voor een goed ontwikkelde technologie zal een enkele gang door deze cyclus van analyse en synthese tot een bevredigend resultaat leiden. In technologisch onderzoek zijn meestal meerdere cycli nodig om te corrigeren voor onvoorziene verschillen tussen de experimentele resultaten en de theoretische voorspellingen.

In dit proefschrift is de hierboven omschreven methode toegepast op een aantal nieuwe componenten op het gebied van de geïntegreerde optica: een nieuw type richtingskoppelaar, polarisatiesplitsers en golflengtedemultiplexers gebaseerd op een nieuw concept, de optische *phased array*, en een nieuwe techniek voor het koppelen van glasvezels aan optische chips. Deze samengestelde componenten kunnen worden beschreven als een combinatie van drie elementaire componenten: golfgeleiders[1], vrij ruimte[2] en juncties[3]. In het proefschrift is de werking van deze componenten beschreven uitgaande van de bekende werking van de elementaire componenten. Verder zijn er ontwerpmethoden ontwikkeld en experimenteel getoetst.

Het in dit proefschrift beschreven werk vormt een stap in een voortgaande ontwikkeling die moet leiden tot een operationeel toepasbare technologie. Het experimentele werk is uitgevoerd met behulp van aluminiumoxyde op siliciumsubstraten. Deze technologie geeft uitstekend reproduceerbare resultaten, hetgeen voor het onderzoek van groot belang is, maar is niet erg geschikt voor het vervaardigen van meer gecompliceerde optische chips. Een volgende stap, waarmee reeds enige resultaten zijn geboekt, is het toepassen

1) Zowel recht als gebogen.
2) Drie-dimensionaal en twee-dimensionaal (filmgolfgeleider).
3) Zowel tussen golfgeleiders met verschillende afmetingen, als tussen golfgleiders en vrije ruimte.

van de ontwikkelde concepten in het quaternaire halfgeleidermateriaal InGaAsP (op InP-substraat). In dit materiaal kunnen optische chips, met inbegrip van lasers en detectoren, volledig worden geïntegreerd.

Het perspectief voor een dergelijk programma is goed. De belangstelling voor compacte en reproduceerbare 3-dB koppelaars neemt snel toe. Het belang van golfgeleiders met zeer kleine bochtstralen wordt groter, naarmate er meer componenten op één chip moeten worden geïntegreerd. De behoefte aan optische componenten die geschikt zijn voor toepassing in breedbandige communicatienetten wordt internationaal als een technische uitdaging gezien, zowel op het gebied van coherente systemen als voor systemen gebaseerd op golflengtemultiplexing (High-Density Wavelength-Division Multiplex). In deze laatste kunnen *phased arrays* een belangrijke rol spelen. Hoewel de ontwikkelde concepten primair voor de telecommunicatie zijn ontwikkeld, beperkt hun toepassing zich daar niet toe. Ook in sectoren als optische gegevensverwerking, sensortechnologie alsmede in de consumenten en de professionele sector kunnen ze met voordeel worden toegepast. Met het hier gerapporteerde werk heeft de vakgroep telecommunicatie- en teleobservatietechnologie zich dan ook een goede positie verworven om mee te werken aan de internationale ontwikkeling van de optische technologie.

Literatuur

E. Schuurman, *Techniek en Toekomst*, Van Gorcum & Comp. N.V., Assen, 1972.

Acknowledgements

The results presented in this thesis could only have been achieved through the close cooperation of many colleagues, students, and relatives. I am grateful for their help and contributions.

Professors G. Acket, H. Blok and L. Krul initiated and stimulated the development of integrated optics research at the Delft University. The cooperation of professors B. Verbeek and R. Baets, who are presently in charge of the integrated optics research program, has been particularly appreciated.

Many students contributed to the results presented in this thesis. J. de Groot and W. van Blitterswijk, who I was supposed to coach, taught me the first principles of optical communication technology. Pioneering work on the development and characterization of sputtered aluminum oxide films was done by J. Angenent and Th. Jongeling of the Optics Research Group (Dept. of Applied Physics). W. Pasmooij, M. van Vaalen, P. Van Eijk, P.A. Mandersloot and E.Otto contributed to the development of the prism-coupling measurement technique, which is described in chapter 8. R. Funke and G. Manhoudt performed the first experiments with waveguide bends, thus providing a basis for future achievements in the field of short waveguide bends. E. Pennings further explored the bending problem, both theoretically and experimentally, developed and implemented algorithms for the computation of transmission properties of straight and bent waveguides, which forms the core of the laboratories present modeling software, and discovered the self-imaging properties of Multi-Mode Interference couplers. His work provided Delft University with a leading position in the field of short waveguide bends. P. Schalkwijk and F. Veerman designed and characterized the TMI-couplers and the modified TMI-couplers, respectively, as described in chapter 5. L. Soldano continued work on MMI-couplers and phase-diversity networks, and provided illustrative material. A. Vellekoop performed the experiments with polarization splitters and wavelength demultiplexers based on an optical phased array, as described in chapter 6. R. Hoefnagel and P. Tiggelman elaborated the concept of an aberration-free phased array which forms the core of chapter 6. Ch. van Vliet contributed to the development and the numerical implementation of the bounded-error approach which has been employed in chapter 7. J. Verhoof contributed to the simulations and experiments on the accuracy of ellipsometric measurements reported in chapter 7.

There would have been no thesis without the members of the integrated optics project group. F. van Ham, J. van Uffelen, and A. de Vreede performed many experiments in order to improve the quality of the aluminum oxide films and waveguides. Van Uffelen developed the 2-µm ridge-guide process. De Vreede performed the fiber-mode matching experiments reported in chapter 9. Y.S. Oei, who is in charge of the InGaAsP technology program, contributed to the characterization of the aluminum oxide films and solved many problems in the field of PC soft- and hardware. C. van Dam developed software for optical circuit modeling and generation of multi-layer masks. M. Amersfoort, Ch. de Boer and P. Fransen, although not involved in the aluminum oxide technology, helped in preparing a foundation for future integrated optics research in Delft: InGaAsP technology.

The Department of Electrical Engineering and the Laboratory of Telecommunication and Remote Sensing Technology (TTT) have made large investments in the opto-electronics laboratory. Many members of the laboratory have contributed in solving a variety of problems and helped to build the infrastructure, which provides the technological basis for achievement in the field of integrated optics. The computer capacity, which has been provided for many years by the remote sensing project group is most appreciated. The microwave components group helped in adapting microwave design software for modeling optical circuits. The help of M. Naus and F. Tauritz in preparing and correcting the manuscript was indispensible in order to meet the printers deadline.

From the beginning the integrated optics research program of the TTT-laboratory has been carried out in cooperation with many research groups and institutes. The members of the Optics Research Group and the Optics Polishing Shop (Applied Physics Department) helped in developing the aluminum oxide deposition technique and the prism coupling technique. W. van Etten and J. van Bennekom from Eindhoven University of Technology helped in developing and characterizing the planar phase-diversity network, reported in chapter 5. A. van Veen and K. Bijkerk from IRI[1] measured the desorption spectra of our aluminum oxide films. C. de Haan from the Laboratory of Metallurgy provided TEM-photographs and EDS-analysis results for the aluminum oxide films. G. van Gurp from Philips Research Laboratories provided unpublished data on aluminum oxide film stress and density. P. Lemmens from Philips Research Laboratories polished the end faces of the long wavelength coupling devices reported in chapter 5. G. Krijnen from Twente University developed an improved version of our mask-layout generation program and provided BPM-software used for the calculations on coupled waveguide structures reported in chapter 5. N. Baken, M. Diemeer, and J. van der Tol from PTT research, and the members of the integrated-optics group at FEL-TNO[2] helped through appreciated discussions and exchange of information. J. Kortlandt helped in learning sputtering technique. E. Smit and the other members of the integrated circuits laboratory (IC-atelier) and its successor, the Delft Institute of Microelectronics and Submicron technology (DIMES) helped with the development of fabrication technology. The members of the instruments development workshop (FOW) helped developing equipment and measurement setups.

The help of my wife, Marijke, was indispensible in the period during which the thesis was being written. I intend to make up in the future for the hours

1) Interfaculty Reactor Institute, Delft University of Technology.
2) TNO Physics and Electronics Laboratory (FEL), Waalsdorp, The Hague.

which could not be spent with Adriaan, Marianne and Pieter. I am grateful to my parents, whom I thank for my education, and to my parents-in-law and all the others who helped me to make free the time required for organizing the results of modeling and experiments and writing the manuscript.

Last, but not least, I wish to express my thanks to God, who gave me the opportunity to discover a small part of the immense potential of his creation.

Biography

Meint K. Smit was born in Vlissingen, the Netherlands, on January 5, 1951. He graduated in 1974, cum laude, in Electrical Engineering at the Delft University of Technology. His master's thesis work concerned design and characterization of an S-band microwave amplifier.

From 1974 to 1981 he worked in the field of radar remote sensing, first as a research assistant to the NIWARS[1] and since 1976 as a staff member of the Delft University. He was charged with the development of a ground-based short-range FM-CW scatterometer, the execution of measurement campaigns, and the automation of data collection and processing.

Since 1981 he has worked in the field of optical communication technology with emphasis on integrated optics. He was involved in the development of technology, modeling and characterization of planar optical components in silicon-based aluminum oxide films, and in education in the field of fiber communication technology and integrated optics. Since 1986 he has also been engaged in research on the realization of integrated optical circuits in InGaAsP/InP.

---

1) Netherlands Interdepartmental Working Community for the Applications of Remote Sensing techniques.